THE GENE BUSINESS

EDWARD YOXEN

THE GENE BUSINESS

Who Should Control Biotechnology?

OXFORD UNIVERSITY PRESS
New York

Copyright © 1983 by Edward Yoxen

First published in the United States in 1983 by Harper & Row,
Publishers, Inc.
First issued in paperback in the United States in 1986 by
Oxford University Press, Inc., 200 Madison Avenue,
New York, New York 10016
Reprinted by arrangement with Harper & Row, Publishers, Inc.

Oxford is a registered trademark of Oxford University Press

Grateful acknowledgment is made for permission to reprint the
DNA diagram on page 27, which first appeared in *BioScience*.
Copyright © 1983 by the American Institute of Biological
Sciences. Reprinted by permission.

Library of Congress Cataloging-in-Publication Data
Yoxen, Edward.
The gene business.
Originally published: New York: Harper & Row, 1984, c1983.
Bibliography: p.
Includes index.
1. Biotechnology. 2. Genetic engineering.
3. Biotechnology—Social aspects.
4. Genetic engineering—Social aspects.
I. Title.
TP248.2.Y68 1986 660'.6 86-2527
ISBN 0-19-504042-2 (pbk.)

Printing (last digit): 9 8 7 6 5 4 3 2 1

Printed in the United States of America

Contents

Acknowledgements

Writing a book is, I have found, a lonely business, even though many people take an active interest in its production. Perhaps with the very visible turmoil of a first book, friends and colleagues are particularly concerned to show their support, but I am still surprised by the number of people who helped me along, if only with friendly curiosity about my progress. Some people whom I have to thank had nothing to do directly with the production of this book, but without the help and inspiration in the past of David Dickson, Jonathan King, Jonathan Jones, Susan Wright, Harry Rothman, Jonathan Beckwith, Pauline Marstrand, Jerry Ravetz, Jost Herbig and Jon Harwood I would never have written it. I am also very grateful to Jon Turney, Jeremy Green, Michael Gibbons, Ken Green, Rene Brierley, Gill Greensides and Les Levidow for more immediate assistance. It is also a pleasure to acknowledge my debt to the Crucible Unit of Central Productions Ltd., and particularly to Sandra Sedgbeer, Morag Roberts, Jane Pascoe, Lawrence Moore, Jane Cousins, Julie Sheppard, Liz Millner and Ashley Bruce. I am grateful to the American Association for the Advancement of Science for permission to reproduce the letter by P. Berg and others which appeared in *Science* in 1974. My thanks also to Paul Derbyshire and members of staff of the microbiology department at Warwick University; whilst they bear no responsibility for the contents of the book, their help was appreciated. I would not have survived the gruelling final stages of completing this book without the considerable forbearance, enthusiasm, imagination and humour of Bob Young, the series editor, and Tony Fry, Andy Lowe, Phil Goodall and Michael Green of Collective Design/Midland. Their unstinting commitment to this project was a tremendous source of support. I should also like to thank Caradoc King and Frances Lindley for their help in seeing this book into publication in the United States.

Foreword to the American Edition

by Robert M. Young

Biotechnology is like micro-electronics and new developments in medicine—all are changing very rapidly in ways that affect our experience and the structure of our society as broadly and deeply as the industrial revolution of the eighteenth and nineteenth centuries. Where did these developments come from? Who sets the priorities? If we believe that people should have a voice in the decisions which directly affect their lives, then we have to ask how we can have a voice in setting agendas in these highly expert fields.

As recently as a decade ago molecular biology gave little sense of promise of giving birth to a biotechnological revolution. It was an impressive branch of biology with more than its quota of Nobel prizes, but its practitioners were not seen as potential business tycoons. The same can be said of solid state physics, the very specialised subdiscipline which gave rise to micro-electronics. These two sets of developments from pure to applied science are now likely to transform modernity, as the general public is coming to appreciate. They are already well aware of the dramatic developments in *in vitro* fertilisation, organ transplants and spare parts surgery, where the social issue of setting medical priorities is more obvious and has become part of public debate.

Edward Yoxen is in a good position to write about these matters in biotechnology. He trained as an engineer before turning to questions in the social studies of science, and wrote a thesis about the social impact of molecular biology in the period just before biotechnology began to come on stream. So his vantage point allows him to shed clear light on the social and political issues raised by the developments from prestigious ivory tower molecular biology to venture capital genetic engineering. He has also been involved in various efforts to generate greater public participation in scientific and technological decision-making and has been active in the British Society for Social Responsibility in Science, the Science Technology and Society Association and the Group for Alternative Strategies in Science and Technology while teaching Science and Technology Policy at the University of Manchester.

What distinguishes this book from everything else I've read on the topic is that at every stage he is addressing the social gains and losses, the effects of economic values on academic ones, the tensions between the need for products to make a return on investment on the one hand and the definition of public service and the common good on the other. The result is the best detailed study I know of the social issues raised by new developments in science, technology and medicine.

The road map of his book is a straightforward one. The introduction spells out just how important the wide range of developments falling under the term "biotechnology" are. He then turns to the history of molecular biology which led to the current ferment of ideas, techniques and investments with particular emphasis on the questions which have already arisen about health, safety and public participation. The third chapter is the hard one—he spells out the science of molecular biology and genetic engineering. A lot of effort has gone into making this accessible to lay people, but it may still be worth skimming or skipping until after the rest of the book has been read. . . . The next three chapters make their arguments by taking us through case studies. Edward Yoxen carefully considers how biotechnology is in the process of transforming medicine, the agriculture and food industries, and the production of energy and chemicals. That is, he shows the changes throughout the industries which produce a great many of the products which affect our lives most directly.

What he has been exceptionally good at, in my opinion, is making it plausible that nonspecialists should dare to think about such esoteric matters as monoclonal antibodies, nitrogen fixation and gene splicing. He shows, by example, that the technical matters are accessible enough so that the issues they raise can be part of the public domain. If other writings about new developments in science, technology and medicine can be opened up in the same way, then it is not out of the question that the politics of expertise can be brought out of the ghetto of small coteries in subdisciplines, the research councils and the boardrooms of high-tech chemical and drug companies.

Finding it appropriate to dare to think about the politics of expertise is one big hurdle. An equally daunting one is that a new politics is required if we are to scrutinise and discuss these issues before it is too late, i.e., before they have so much economic and industrial investment behind them that only massive public protest can prevent their deployment. Nuclear power plants and Cruise missiles are relevant analogies. They can be stopped, but it would have been much easier to do so if the research and development phase had been open to public scrutiny and debate. A more plausible candidate for successful public intervention at an earlier stage is the word processor with visual display unit, which is transforming office work. In a society in which the setting of R&D priorities was subject to democratic control, the process of origination of the new office technologies would have involved discussions with office workers. A result might have been less surveillance, pacing and control and less technological unemployment.

The concluding chapter of *The Gene Business* is concerned with the openings which Edward Yoxen believes exist for a greater public role. I'd be more optimistic than he is, but he has chosen to be quite cautious about what can be done in the short run. What is most important, however, is that he shows that a new politics of the role of science in society is credible and important. He has taken a sure-footed step in bringing it into being, and there are many more to take.

It has taken science several centuries to develop its complex language. Indeed there is not a single one. Scientists have evolved so many different technical vocabularies that they can seldom communicate between subdisciplines, much less between different sciences. A public debate demands that these codes must not be taken as sacrosanct. Debate can happen only if new ways of talking about science are developed by people who are determined to break through the barriers which make it seem that science can be isolated from politics. The fact is that politics are occurring on both sides of any communication barrier and that we are all lay people with respect to everything but our particular little corner of expertise. I encourage you, the reader, to join in this task of creating a common culture for the consideration of social priorities and matters of expertise. However you use this book—informally or in teaching—you should feel free to interfere with its text in the name of accessibility. Notes and references are to be found at the end of the book.

1

The Life Industry

I never thought that I would live through a revolution, but I believe that there is one going on now. No Winter Palace has been seized, no Bastille stormed, no monarchy abolished. Political orders around the world remain, alas, very much the same. But on the industrial and scientific fronts major changes are gathering momentum. A technological assault is being prepared that will transform the economies of developed and developing nations. Its substance is the engineering of life processes for commercial ends: biotechnology.

These are early days still. Much remains to be fought out. The deeper recesses of the economy and the cultural hinterland have yet to be opened up. But it will happen. However, unlike some of the social transformations just mentioned, where the control of the state was seized from the faltering hands of a dominant class, this revolution is being planned from boardrooms and government ministries. It is a revolution by and for corporate capital. It is a rebuilding of industries, companies, universities and laboratories to sustain the present mode of production. Making the revolution means managing and carrying out genetic engineering research.

Biotechnology has an ancient lineage. It is as old as the first fermented drink, the first bowl of yoghurt or the first piece of cheese. For thousands of years people from many cultures have used biological processes, in a controlled fashion, to make foods, dyes, drugs, fuels, adhesives, paper and fertilisers. Many of these procedures lie at the heart of traditional industries like brewing or the manufacture of dairy products.

Into this complex of traditional crafts, household routines and long established industrial practices has now come the dynamism of high

1

technology and advanced laboratory research. Not only are traditional micro-organisms like moulds and yeasts being geared up to much higher levels of productivity, but new organisms are being built to order to carry out all kinds of hitherto inconceivable tasks. Bacteria can now make human proteins, excrete plastic, manufacture anti-freeze, digest wood chips and turn them into edible protein, live on waste oil, break down herbicides like 2,4,5-T, extract metals from their ores or accumulate them from sea water, and turn human sewage into food. Add to that what yeasts can do (principally, they can make alcohols from all kinds of improbable materials), what moulds can do (which is a great deal more than just putting the blue veins in Roquefort), what cultured plant, animal and human cells can do, and you have the makings of a revolution.

In a mere seven or eight years biotechnology has grown from being an esoteric research topic of university scientists into the foundations of a new industrial movement, a new wave of investment, commercialisation and production. Many of the things that were discussed as science fiction five years ago have already happened. This is not just a change of technique, it is a new way of seeing. It is now possible to think of making organisms to a specification to carry out particular industrial tasks. The limitations of species can be transcended by splicing organisms, combining functions, dovetailing abilities and linking together chains of properties. The living world can now be viewed as a vast organic Lego kit inviting combination, hybridisation and continual rebuilding. Life is manipulability.

To the participants, mainly scientists and business executives, what is happening is exciting, unexpected, novel, demanding and lucrative. So far there have been very few casualties, only the odd minor bankruptcy, one or two lawsuits settled out of court, some acrimony about sharp practice and plagiarism, but nothing catastrophic. Most of the people involved are having a ball, and some of them have made sizeable stock-market fortunes, with the prospect of more to come.

Yet already the first wave of technological optimism is past. The first round of investment has happened, and a whole series of small companies have been formed, many of which are now finding the going very hard. Succeeding in this business is not just a matter of having smart ideas about what to do with genes. It is also a matter of keeping the money flowing in before any products appear on the market. There is now a pause for reflection in academic and business circles in which many will ponder whether it is really worth investing time, energy, money and status in projects that only a year ago appeared very appealing. With this book I want to exploit to the full the opportunities of that change of mood and to raise questions about the relations between the manipulability of nature and social manipulation.

The Obsolescent Breast

For example take human milk. You may remember a scene from the film *Sunday Bloody Sunday:* not the more sensational one of Peter Finch and Murray Head making love, but the one where Glenda Jackson and Head, while babysitting in a trendy household for the weekend, pull out from the fridge a milk bottle with some whitish fluid in it. Just before it goes over the breakfast cornflakes one of the children says, "That's Mummy's," and the breast milk goes back in the fridge. But in years to come human milk might be found on supermarket shelves, derived not from women, but from bacteria.

Breast milk is an extremely subtle combination of fats, proteins, peptides (mini-proteins) and antibodies, and every species has its own characteristic mixture. Each of the different components has a function, some of them nutritional, some conferring protection against disease, some aiding the baby's digestion. What some of them do can only be a matter for speculation at present. Nobody is likely to come anywhere near a good imitation of human breast milk, not least since its composition varies during the feed. But the makers of breast milk substitutes do certainly try.

Scientists can now get some of the components of milk made in bacteria by genetic engineering. The idea is partly to study how the synthesis of the constituents of milk is controlled. How, they ask, do the hormones involved in lactation control the production of milk? But this kind of work is also of interest to companies like Nestlé that make milk substitutes. These have proven very controversial products when sold in Third World countries. Conceivably such knowledge might be put to work in the design of new, even more appealing milk powder. What this would mean for infant nutrition, were it ever to come about, remains to be seen. It could be the "greatest" thing since sliced white bread, with all the ambiguities thereof.

This is just one instance of the power of biotechnology to take a bodily substance, like the components of blood, sweat or tears, and to make them artificially in reprogrammed bacteria and in enormous quantities. Artificial human milk would be one instance in which biotechnology could alter a whole set of symbolic and cultural relationships that people have with their bodies. Bodies, organs, glands and tissues are becoming irrelevant and obsolete, because bacteria reprogrammed to act as incredibly productive chemical factories can do the job better. How this is done we will consider later. Putting microorganisms to work in this way has so far only been done for a few score of chemicals. But in principle *any* molecule made by *any* cellular organim, be it sperm whale, pyrethrum plant or human being, can now be "grown up" for sale in vast quantities. The scale of what's possible is astonishing.

Who Needs Plants?

For some reason this came home to me only recently when I was reading a magazine produced by radical midwives. One brief article praised the virtues of raspberry leaf tea for pregnant women. The leaves contain a substance called fragine, which acts as a muscle tonic and is particularly effective in strengthening the muscles of the pelvic floor. Presumably it is a traditional brew, once well known, now lost from the cultural memory of women disciplined by hospital childbirth.

Suddenly it struck me that substances like fragine were obvious candidates for biotechnological manufacture. At the moment fragine is concentrated in raspberry leaves as they grow. Molecules of fragine are constructed in the leaf cells as part of the business of being a raspberry bush. But in principle there is no obstacle to isolating the system responsible for this process and transferring it to bacteria. By this means the microbes would be reprogrammed to start making fragine. It is now possible to break some plants down into single cells that will still function in the disaggregated state, that is, when split up into individual units rather than locked together as a plant. Although this is a hypothetical example, the point is that you could make fragine in an organism other than a raspberry bush.

Now it may be that fragine is not that easy to produce except in a raspberry leaf; or it may be that its efficacy is actually not very great; or it may be that there is no money in developing this process. But whatever the practicalities, economics and clinical advantages of making fragine without a plant might be, the symbolic loss would be real. A herbal tea, a cheap potion from the subculture of traditional medicine, would be turned into just another pill to be prescribed by the doctor.

There can be problems with traditional medicines. The dose can vary, there may be impurities in them, there may be medical circumstances under which they should not be taken. However, in this case we are talking about transferring a substance from a context of trust and mutual support to the alienating world of modern medicine. Chemically it is the same thing, but the meaning of its use would be completely transformed.

Biotechnology, then, offers that kind of power. It allows us to sift through the plant kingdom, looking for profitable substances to be made in vats of green goulash, or in tubs of microbes living in industrial fermenters. The fragine example may be a silly one to pick because it is probably so marginal economically. Nobody makes a killing out of dried raspberry leaves. But the economics of quinine, tobacco or heroin are a different matter.

For the Indian grower used to selling cinchona bark to be turned into quinine, the new technology could be a disaster. As far as tobacco is concerned, cigarettes have already been made from cultured plant cells produced in a vat. The economics of growing tobacco without plants are not yet competitive, but in time perhaps all that land which the tobacco giants control will be released for other things. You can also get opiates from plant cell cultures, and some producers in this sector have well deserved reputations as less than wholesome people. What could be more profitable than to invest in the latest biotechnology and disguise the factory as an olive oil business?

If all these things could be produced without the trouble of ploughing, sowing, watering, nurturing, harvesting and shipping the product overseas (and there is research that suggests that they can) then why waste time with a plant?

Steak from Chips

Many of the industrial processes that constitute urban industrial civilisation make a mess. Papermaking, spinning, forestry, making sugar from cane, sweetmaking, pig-rearing and countless other industries generate waste material, sometimes on a gargantuan scale. Sometimes it is burnt, sometimes pumped into the sewage system; sometimes it is left on the land; occasionally it is recycled.

There is now the possibility of turning some of the detritus of our technological society into food. The ploy is to use waste material as a food source for bacteria, which break it down, and turn it into protein. The protein-laden microbes can then be dried, mulched, turned into pellets or beaten into pleasing shapes, and the resulting product is edible, if you like that sort of thing. No doubt you could add some flavourings and put it into sausages and no one would know the difference. Professor Moo-Young of Waterloo University in Canada has developed a process that can turn forestry waste, bark, twigs, sawdust, wood chips and trees too small for the sawmill into a protein-rich food.

The big question seems to be whether it is worth the trouble. Generally speaking, commercial lumber companies could not care less about the devastation they leave behind them. They gut forests with bulldozers and tree-movers, strip-mine the saleable timber and leave thousands of acres of helter-skelter rubbish and violated ecosystems. To stop gulping, and tidy up, would only eat into profit margins. So for Moo-Young's protein to end up in hamburgers, the economics of recycling would have to change. But prospects are good; he has sold the rights, and several governments are scaling up the process to commercial levels.

What this means is that either the appeal of processing wastes in this way would have to be enhanced, say by a massive increase in the cost of meat; or that the costs of not doing so would have to increase, say by government requirements to restrict the way in which forests are stripped. Moo-Young claims that we are almost at that point. Although those who run timber companies tend to be wild and woolly believers in free enterprise, backwoodsmen, if you will, there is no doubt that some governments could twist their arms. With the prospect of food from wood waste such silvan rapine becomes doubly unjustified.

That is not the only waste material on which we could go to work. Human sewage contains forty per cent usable protein, yet we simply pour it away. From one point of view, doing anything other than jettisoning it after bacteria have broken down its odoriferous and dangerous components is distasteful. Yet that option is in fact a luxury, or an indulgence. Can we in the developed countries continue to waste this resource, because we have acquired the wealth to make protein in different and extremely inefficient ways?

In less developed countries other materials are also available as part of the legacy of colonialism. There are wastes from growing and processing sugar, coffee and oilseeds, which cause enormous problems at the moment, adding environmental insult to the injuries of continuing economic exploitation. These could become an asset. That is to say, they could be used to grow microbial food, even though the immediate economic benefits would go to those who own the sugar, coffee and oilseed plantations.

At the moment new bacterial and fungal foods are being grown on nonwaste inorganic materials. ICI (Imperial Chemical Industries) has a process that uses a kind of alcohol made from North Sea gas, ammonia and air as the gruel on which a special bacterium will grow. The resulting culture is turned into a food called Pruteen for pigs, cattle and hens, as an alternative to soya meal. ICI scientists talk of licensing the process, so that small feedstuff manufacturers in Britain can process local wastes. The technology could also be exported to countries with natural gas such as Mexico and the Arab Gulf states. Rank Hovis McDougall have their own bug, a mould, which can be eaten by human beings. Volunteers in their works canteen have already done so; it is said to taste of mushrooms.

Biotechnologies are being developed to turn all kinds of offal, garbage, slime, slurry, sludge, waste and by-products into a basic food substance. Mixed with flavourings, plasticizers, binding agents, preservatives and dyes, they could have a great future. Biotechnology may yet give us a fish finger made at the back of the paper mill, or a meat pie from a sugar plantation. If it turns out that the bacterial pap is too rich in toxic nucleic acids for direct human consumption (and that is the problem with ICI's Purteen) then it can always be fed to animals to get them to convert it into meat, the most favoured form of protein in the Western diet. Let them eat bacterial cake. The world might be a cleaner place for it.

Moving Genes Around

Most of the examples I've mentioned so far have concerned a process of artificially making biological things that has been revved up, reoriented or even specially created by moving genes around. One *can* find microbes in nature that will break down spilt oil, live in boiling acid, accumulate uranium, cadmium or copper in their cellular matrices or break down pesticides. The range of already existing bacterial skills is amazing.

But the essence of biotechnology is to try to improve on this by combining properties, often from radically different species. ICI found its Pruteen micro-organism on a playing field. They chose to supercharge the micro-organism genetically by splicing in some new genes. This trick involves moving genes around, taking particular traits developed way back in the evolutionary career of one organism and building them into a new one, put together to a novel specification. Making human breast milk in microbes is the result of a far more radical combination. Bacteria do not have mammary glands, and they do not secrete milk proteins, but scientists can make them do that by introducing the necessary genetic instructions from human cells.

The examples of plant cell biotechnology to produce such things as quinine might seem not to fit the pattern, because cinchona cells growing in culture will make quinine without any extraneous genetic information. In the right environment they just get on with the job, without needing to be part of a plant. However, plant cells in that state of disaggregation can also be given new genes. Alternatively they can be fused with cells from other species. The resulting hybrids will indeed grow up into fine upstanding plants. These kinds of techniques allow you to make a "pomato" (potato-tomato). There is even a plant–human cell hybrid. It is not a walking, talking, living daisy. It is a plant that in addition to being a plant makes a number of human proteins. The aim is to boost its nutritional value.

The ability to cut and splice in this way, to make vast leaps across millions of years of evolutionary divergence in order to construct a life form to order, is very recent. Although plant and animal breeders have imposed their will on nature for several thousand years, to create the cattle, chickens, potatoes, wheat, maize, nectarines and blackcurrants we recognise today, these radical new combinations of genes are at most only ten years old, and many of them less than that. Biotechnology, as we understand it today, depends on the ability to isolate gene molecules, to get them transported from cell to cell, and to get them reintegrated into the cellular machinery in exactly the right place, so that the new system churns out a totally new molecule. Biotechnology now depends on moving genes around. The ability to do this is a skill that commands a high price in the markets for biological labour.

Nowhere is the power of this ability better demonstrated than in the rapid progress being made with human genetic engineering. Five years ago it was customary for biologists to dismiss the idea of human gene manipulation as wild, unsubstantiated, irresponsible speculation about things that might happen far in the future. In 1980 there was one controversial, unsuccessful attempt to use gene splicing to correct a genetic defect in two people. Before this book appears in print it is very likely that there will have been others, because of the amount of effort being devoted to refining the techniques. In 1981, researchers at Ohio University and the Jackson Laboratory in Maine succeeded in getting rabbit globin genes—genes that control the production of one of the constitutents of red blood cells in rabbits—incorporated into fertilised mouse eggs in such a way that in some of the resulting mice, rabbit globin could be detected, and the rabbit gene was found to have been passed on to some of the next generation.

That may sound pretty boring, but to achieve this movement of a gene from one species to another and get it passed on to the next generation is technically amazing. Whatever the goals of the researchers concerned may have been, these are the kind of skills one needs to do human genetic engineering, or, come to that, to do cow, pig, hen, horse, sheep or rabbit genetic engineering. When you consider that the international trade in frozen pedigree cow embryos, which are reimplanted in other cows, is now worth millions of dollars, it may well be that animal genetic engineering will come on stream much sooner than the correction of human genetic defects.

Genetics now offers an exceedingly powerful handle on nature. Until recently we have had to be content with the slow and painstaking processes of plant and animal breeding. The longer the generation time, the longer the period before new varieties appeared; and, of course, only combinations of organisms within the same species, or slightly further apart, could be contemplated. Yet now we can take a human gene and put it into a bacterium, or mix up rabbit genes with mice genes, without chaos. This is an amazing degree of virtuosity, a novel form of power that is awesome in its implications.

The theme of moving genes around runs throughout this book. I dwell on it repeatedly in an attempt to unify the analysis, to draw out common features in what may otherwise seem very disparate fields of endeavour, and to highlight the change in our relation to nature that biotechnology embodies. Even though some of its practitioners will deny it, biotechnology is a gestalt switch, an electrifying change of perception, once you see what is possible.

Technological Revolution and Social Dislocation

These examples suggest, I hope, something of the scale, pace and transformative power of biotechnological research. I am scarcely the first to invoke the analogy of a revolution to convey the dimensions and texture of what is happening. A recent UK government report put it like this:

> Genetic manipulation . . . has become a practical and quite general proposition . . . This advance in our view confers on biotechnology an importance comparable to that of atomic physics, electronics and most recently, microelectronics. It has been said that "biology will launch an industry as

characteristic of the twenty-first century as those based on physics and chemistry have been of the twentieth century."

Even allowing for some exaggeration, for some rhetorical flourishes, for the enthusiasm of entrepreneurial promotion, it is clear that major industrial developments are in train. What counts as waste, what gets used as food, what defines a species, how procreation occurs, whether living things can be patented and owned—all these issues are being reassessed by the capitalisation of biotechnology.

Yet if the emerging group of technologies generates a transformation in production comparable with, say, the advent of mass production in the nineteenth century, then we can expect massive social dislocations: changes in the uses of land, in the balances of international trade, in technological monopolies and dependences, and in the value of raw materials. Mass production brought with it changes in class structure, unemployment for skilled workers, an increase in the pace of work, a shift in control at the point of production, and major alterations in patterns and forms of consumption. It produced losses *and* gains for workers and consumers. Throughout this book I shall be trying to show what the losses and gains are likely to be from the take-up of biotechnology in different industrial sectors.

There are no easy and clear-cut predictions that one can make. If I give the impression of ambivalence sometimes, that is because it is difficult to trade off conflicting considerations and come to a definite conclusion. In any case, I am not trying to lay down a line on this or that product, and I am certainly not trying to damn biotechnology wholesale. I want instead to lay out the issues, to excavate under the rhetoric and point to implicit choices that have already been made, and to get the social implications of biotechnology more widely discussed. The magnitude of what is coming down the pipeline of innovation is such that the whole genus of technologies that we label "biotechnology," that "family" of new ways of making things, demands a thorough scrutiny. It needs a social audit prior to implementation, a "technology assessment" according to particular economic and political assumptions, a public weighing of probable losses and gains—an informed debate, with sufficient strength to enforce a change of strategy, if that seems desirable. Revolutions, do, after, all, change direction sometimes: they may capsize and founder; sometimes they are hijacked; sometimes they go horrifyingly astray; sometimes they are steered into new and more liberating channels.

The Biotechnological "Agenda"

At this stage nothing is certain, although the general pattern of change is fairly clear. I find it helpful to think in terms of an agenda for biotechnology; indeed that idea helped to organise this book. I shall be moving from things that are happening now to things that lie some way in the future.

The pharmaceutical industry is already putting genetic engineering products on the market. Contemporary high-technology "curative" medicine has proved a very profitable domain for drug companies, many of which have only achieved their present opulence in the postwar period. Their problem now is to find new products that can then be sold in large quantities, with a large profit margin, to recover the massive costs of development and promotion.

One tactic is to take medicinal substances that can be extracted from human or mammalian organs at some expense such as insulin for diabetics, or interferon for cancer research, and to make them more cheaply in bacteria. Another is to use genetic engineering to increase the yields of antibiotics from moulds that produce them. Another is to screen traditional herbal medicines for unexploited natural products. Yet another is to concentrate on veterinary products where sales are massive and recurrent, since the "patients" tend to get slaughtered. Indeed the first product to go on the market is a vaccine that is supposed to prevent diarrhoea in pigs.

But overall the name of the game is to foster the demand for new drugs, to concentrate on high-value products, where competition can be minimised, in order to recover the huge development costs. Genetic engineering is the key to staying in that kind of business. Pharmaceutical companies realised this in the early 1970s, several years before scientists grasped what they could do for the big drug houses. Myopia would be a serious defect in a multinational corporation. The results of that forward planning are coming on to the market now. They serve to ensure that health is sold as a commodity.

In the case of the food industry new developments are also just appearing. I have already mentioned single-cell protein from ICI and Rank Hovis McDougall. The German chemical company, Hoechst, is in that business too; BP used to be but was driven out. These are major projects, that either service the production of a high-cost, high-status, resource-intensive food—meat—or seek to supplant it. Less spectacular developments in this area are new sweeteners, like high-fructose corn syrup, which is big in the United States and said by some to be a way of attacking Cuba by driving down world demand for sugar. Other products include G. D. Searle's new sweetener, Aspartame. Tate and Lyle has its own project, making a sweetener called Talin in bacteria. In all of these the tendency is to put capital into chemical plant rather than into land and agricultural labour.

In agriculture, where developments are obviously linked to what happens in the food industry (as any East Anglian pea-grower will tell you), major changes lie a little further in the future. The bulk of the work is going into producing plants for high-technology, mechanised, energy-intensive agriculture. Sophisticated science lies behind the development of new hybrids, like triticale, a cross between wheat and rye. New wheats are being bred that will grow in salty soils. One longer term goal is to produce cereals that will fix their own atmospheric nitrogen, removing the need for expensive artificial fertilisers. The chemical companies that make the fertilisers are anticipating this event by developing bacterial products that will do the same job, or by planning to sell the seeds of these remarkable new plants when they appear. It is also just possible that genetic engineering will be used to raise yields in meat or dairy farming. Since even supercows must excrete, it is likely that farm production of methane from animal manure will become more common and that new kinds of microbes will produce the gas.

In the chemical industry there is every sign of crisis, linked to costs passed on from the oil industry and to the decline in world trade. Biotechnology will be important in the longer term to the surviving corporations, which is one reason why it is the chemical companies that are investing on a spectacular scale in university research, with a view to getting a return in a few years time.

Many of the deals with university departments concern medical products, which represent one possible path of diversification. Another option open to these concerns in the longer term is to take a feedstock, a starting material containing carbon and hydrogen in simple combinations, and to turn it into something new. This is ICI's strategy with Pruteen. Also they can try to substitute for existing products like fertilisers or textiles. ICI has a bacterium that overproduces a chemical that can be spun into a fibre. At the moment they think it could be used for surgical twine; but if they keep at it then perhaps in time it could be turned into a reasonable shirt. More radically, new cheap starting materials, elaborated into whole families of chemicals, have to be created to replace oil as it becomes scarcer and more expensive. At the moment nothing looks very promising for an industry which was used to beginning with cheap and simple hydro-carbon molecules and turning them step by step into a myriad of complex chemical products.

Designing New Futures into Nature

There is a lot of money riding on the outcome of research now in progress, which means that corporate investors will be very deter-mined to see that their plans for new pharmaceutical products, new agricultural plants and new sources of supply for the chemical industry are not derailed.

But nothing is inevitable. The view of technological change held by many people is that it just goes on happening relentlessly, except when Luddites intervene and wreck things. However, I believe that new technologies, processes and products have to be dreamt, argued, battled, willed, cajoled and negotiated into existence. They arise through endless rounds of conjecture, experiment, persuasion, ap-praisal and promotion. They emerge from chains of activity, in which at many points their form and existence is in jeopardy. There is no unstoppable process that brings inventions to the market. They are realised only as survivors.

If this view is correct, then the scale, pace and social impact of the biotechnological revolution must be open to negotiation. There must be alternative pathways, which exploit the present scientific possibili-ties, but which frame them differently. Those alternatives should be capable of realisation, but only if enough people come to see their value and fight for them. This book is written in the service of that idea, to break up the mystifying rhetoric that sells everything as prog-ress, and to help us imagine alternative futures.

There is, meanwhile, a school of thought that says that biotech-nology is nothing new, so that no one ought to get excited. It is, we are told, no great change in industrial production, no quantum jump into a new mode of manufacture. It is as old as the fermentation of alcohol, or cheese-making, or baking, or the making of compost. It is a set of crafts hallowed by tradition, broadened in recent centuries into a set of industries. In this view, biotechnology is rooted in domes-tic skills and folk memory, familiar as the basis of established indus-tries, already practised by a cadre of reliable, experienced, decent, down-to-earth technologists and applied scientists, who churn out sta-ples and basics for satisfied consumers. The example of centuries of vinegar production should help us to sleep easy in our beds. Sensational talk of genetic engineering, cloning, patenting life forms, cancer epidemics and tampering with evolution is but froth stirred up by newcomers—geneticists with extravagant notions of their value to industry and irresponsible journalists bent on confusing a gullible public. Of course, any competent biotechnologist ought not to be worried by froth.

There is something in this view. Fermentation is an ancient art, that long ago was built up to an industrial scale. Traditional specialists are being swept aside and their place usurped by scientists from high-status academic fields, whose skills may not be broad or robust enough to cope with the exigencies of industrial production. But what makes contemporary biotechnology so distinctive is the scale of the changes in view, the encompassing breadth of the turn to biological processes, and the almost universal reliance on applied genetics to create the new developments.

The ability to move genes between organisms, to reprogramme one organism with hereditary instructions taken from another, is central to the present regrouping of industrial forces. This is an idea to which we return again and again in this book: that it is the controlled transfer of genetic information into a host organism, be it a bacterium, a yeast, a plant or a mammal, that is revolutionising industrial production. And that it is not just industry that will be transformed. This is also to be a cultural revolution, a shifting of conventions and attitudes. The restructuring of the industrial base, catalyzed by biotechnology, is changing what the very word *life* means. The boundaries, compartments, origins, limits and potential of the living state are being remade by molecular genetics. Biotechnology is the projection on to an industrial scale of a new view of nature as programmed matter.

Every age constructs a model of the living world, built up from the theories, and the social and political imagery of the day, that highlights or emphasises particular aspects of our understanding. In the eighteenth century, an age of classification in botany and zoology, the emphasis was on harmony and systemic order. Nature was a catalogue of organic forms, each fashioned by an ingenious creator, each with a place on a Chain of Being that stretched from inanimate matter to God. The scientist's task, confronted with this majestic scheme, was to classify its elements, to contemplate the subtlety of the connections that held it together and to reveal the harmonious functioning of particular parts. In the nineteenth century, the picture changed with the idea of dynamic, evolutionary change, based on competition and struggle. "Nature red in tooth and claw" was the image for a new age of rapid industrialisation, aggressive business practices, and intensifying struggles between capital and labour. Organisms were approached in a different light as the products not of design, but of millennia of competition with other species, in which the better adapted eventually outbreed their competitors.

The dominant image of nature in the second half of the twentieth century, deepened by the insights of genetics, is less reverential than that of the eighteenth, and places less emphasis on struggle and competition than that of the nineteenth. Nature is a system of systems. Organisms function, reproduce and evolve as systems ordered by their genes, "managed" by the programme in their DNA. Life is the processing of information. The same concepts, drawn from computer science, cryptography, programming and control engineering, fit the design of an early warning system for intercontinental missiles, the patterns of activity in an anthill, the use of a numerically controlled machine tool, the control of blood pressure and the way in which cells make protein molecules.

We are now at the stage where merely thinking of organisms as programmed systems is giving way to the activity of reprogramming them. Scientists can now intervene in nature, constructing to order, as a microchip designer might decide what functions to realise in a piece of silicon, or a computer engineer select a range of modules with which to build a data-processing system. The analogy is not trivial. As microbiology becomes industrialised as biotechnology, that kind of construction activity, which has already shown its prodigious potential in micro-electronics, computing, robotics and systems engineering, will take the foreground in the life sciences. As in the field of inanimate hardware, it is astonishing what becomes possible when you start combining modules and functions.

Thus our image of nature is coming more and more to emphasise human intervention through a process of design. More and more, genes, organisms, biochemical pathways and industrial bioreactors and processes can be realised according to a prior specification. Now, the essence of life is its constructability.

We stand at a fascinating and pivotal moment in the history of technology, production and the life sciences. Access to tremendous manipulative powers in biology is being opened up and we would have to be particularly insensitive not to be awed and excited by what is happening. It is not that often that sciences are in ferment, and even when it happens it is rare that it shakes public consciousness, has an immediate technological payoff, and causes major waves of legal, political, moral and economic debate.

With biotechnology all these things have come together. Not only are there major conceptual shifts going on that will trickle down into the textbooks at some point, but the prospect of massive industrial change is imminent, various epochal legal issues occupy a central position, and complex political questions bearing on scientists' responsibility to society have to be debated and resolved.

Before we get to grips with what is happening in various areas of biotechnology, I want first to suggest a way of thinking about what is going on, and then to give a rapid introduction to some ideas about biology, which inform the technical vocabulary of the book. These concerns take up the next two chapters.

2

How Life Acquired
a New Meaning

Artists sometimes begin a picture by blocking in the major areas of colour, just to see how these will work together in the final construction. I am doing something similar here. The point I want to get across is that the three themes introduced in this chapter fit together. It is their interaction and mutual connection that is important. Each theme takes a sentence to state. A particular view of nature has helped a scientific élite to come into existence. That group conducted a brief but valuable experiment in public participation in science in the late 1970s, by calling a temporary halt to their research, despite its evident promise, and thinking aloud in public about whether it was safe to proceed. When the size of the technological payoff from this new research area of gene-splicing became apparent, and the passions stirred up by public debate looked disconcertingly powerful, the experiment in consultation was wound up and the scientists concerned threw all their energies into the formation of a new industry based on biotechnology. Let's run that by again: technical virtuosity, developed over decades of selective funding to transform the life sciences, opened up a new frontier. On the threshold some of the people involved held back to consider what might be coming next, conscious that they would not be able to contain themselves for long. After this moment of hesitation, a wave of investment in biotechnology has swept forward, taking virtually all of the scientists with it. They now find themselves in an industrial revolution, taking the role of highly paid technicians, drawing on ever increasing skills, with less and less inclination now to ask what will come of all this. The obvious casualties of this changed atmosphere are public participation and scientific accountability. After a brief interlude, the scientists and businessmen are again in full charge and are assuring us that we are in good hands.

The Meccano View of Nature

Somewhere among the piles of notes from my student days there is a preamble for a course on cell biology, which says something like, "Organisms are self-assembling, self-maintaining, self-reproducing machines that operate at room temperature and pressure." I did not quarrel with that statement at the time, nor do I now. It is an accurate and productive way of thinking about living things. It encapsulates an attitude to biological organisation and function that has enabled scientists to analyse the working of cells in immense detail and to focus intently on particular processes, like that of heredity, which have proved to be amenable to this way of thinking. But I do remember pondering again and again the peculiar deadness of those lines. Somehow this mechanical attitude seems to end up all too often in piles of data about some feature of the cellular machine, obtained by busting it into its constitutent molecules to find out what they are and how they fit together, without sufficient sense of the intricacy or delicacy of the whole thing.

This basic and pragmatic simplification that says, "Let's treat any organism as if it were a rather complex kind of machine" has been immensely enriched over the last thirty-five years by the notion of "information." In a sense the decisive, energising perception of biology since the Second World War, the key to its strength and vigour, has been one that treats organisms as information-processing machines. They begin as packets of information; they organise themselves through a process of programmed self-assembly; they operate on their environment in a controlled manner according to genetic instructions; they reproduce by condensing their structure and functional coherence into a transmittable form—that carries a message containing the instructions in a code that organisms can "read." To think of life in this vocabulary is basic to modern biology.

This conceptual framework organises a great deal of research, though not everything. Among other things, it suggests an immense interlocking series of minor projects to be tackled, which are all concerned with the role of particular molecules in an informational system. All too often, amazement at the intricacy and evolved complexity of living things seems to get lost in the analysis of fine structural detail, such as how a particular molecule has been put together as a string of chemical units. Biologists' attention seems riveted by experiments which are intended to get particular molecules to reveal themselves.

A common technique is electrophoresis, where tiny fragments of once living organisms are forced to race each other through jelly to grade themselves by size. How big is this fragment? What is its charge? Does it have any subunits? How much information is needed to put it together? How does it lock on to other molecules? Questions like these are the stock-in-trade of an area of the life sciences called molecular biology.

The main preoccupation of molecular biology is the analysis of how genes direct living systems. For molecular biologists, life is what genes do. For them genes are the key to life, and one need look no further than this for the central problems of biology. In their hands biology has become a kind of flatland in which the only activity is the processing and transmission of genetic information.

Although this conceptual move is not to everyone's taste, in scientific terms it has proved amazingly productive. It is an analytic framework that is also particularly well suited to the internal economy of science, where salient facts can be exchanged for career status and resources.

In a sense it is a remarkably *un*biological biology. It is a science that draws its strength from an abstract and desiccated view of life. Organisms are merely systems, and can be studied as systems, reducible in the last instance to a particular kind of logic. The study of process and form has been given up for the analysis of structure and linear order. It is a view of living things far removed from everyday experience and the aesthetic appreciation of shape, elegance or anatomical subtlety, and equally far from naturalists' understanding of lifestyle and habitat. It is a biology built on fundamental abstractions such as the notion of a universal code, the idea of information, the metaphor of a programme controlling cellular activity.

One biochemist, dismayed by the lack of reverence that molecular biologists seem to display before the entities they strip apart and grind to bits—as do biochemists too, by the way—has called it the "strip-mining of nature." That may capture the rapacity of the enterprise, the desperation to keep up the rate of production of facts and the long hours worked with powerful machinery, and insufficient attention to the overall effect on the natural landscape. But it does not allow that occasionally all the destructive, analytic labour produces unifying insights of great beauty—and manifestly, as I shall make clear, molecular biology has done that. Nor does it make any reference to the notion of information, or the fact that now molecular biologists are engaged in synthesis and construction, putting new organisms together.

I prefer to think of molecular biology as the expression of a Meccano view of nature. With a fairly simple conceptual kit and with a limited number of elements, molecular biologists have been able to represent living nature with a series of increasingly complex mechanical models. They have spent years figuring out what pieces there are in nature's Meccano set, and how they fit together. Some of the more theoretically inclined have even examined the very principles of construction, the rules of order and geometry built into the Meccano parts. And now, finally, since the early 1970s they have figured out how to start bolting pieces together, making new models that are not even in the instruction books. To push the analogy to its facetious conclusion, we could say that molecular biologists are now realising what they can build and just how pleased their new patrons will be with their inventiveness.

Lest the use of this analogy backfire on me by trivialising molecular biology, I should also say that this kind of model-building represents an extraordinary virtuosity in teasing apart the components of biological organisms and forcing them to reveal the detail of their construction. It is the product of decades of accumulated expertise. It is the achievement of an acclaimed scientific élite, the result of intensive labour and the embodiment of carefully nurtured skills in argument and at the lab bench.

Molecular Biology: Coining the Term

This kind of biology and its strategic simplifications have their roots in the 1930s. It was at the end of that decade that the term "molecular biology" was coined and stuck as a label on certain avant-garde research projects. This innovation, shifting attention to the molecular level of organisation, was no accident, no unwilled bonanza from a technical breakthrough. It was the result of a series of decisions at the Rockefeller Foundation, an immensely influential philanthropic agency, that deliberately set out to create a new kind of biology. Without that, molecular biology would still have emerged, but at a later date and in a less organised form.

By the turn of the twentieth century, John D. Rockefeller Sr., like other American industrial magnates of the time, such as Mellon, Carnegie, Vanderbilt and Morgan, had accumulated a vast fortune, through his ruthless endeavour in the oil industry. He was Standard

Oil, later broken up into Esso, Humble, Sohio, Socal etc., and latterly, Exxon. His desire to moderate some of the considerable hostility being shown to his vast business complex and a conviction, shared by the so-called Progressive Movement in the United States, that intelligent, carefully directed philanthropy could reform and reinforce the institutions of industrial society, led him to order the creation of various charitable trusts.

His goal was to promote the "rationalisation" of all features of urban, industrial, capitalist society, to "naturalise" its all too evident inequalities and to banish the spectre of socialism as a source of alternatives. To this end he was prepared to spend enormous sums of money to strengthen a reformed economic and political system. This was not mere propaganda; it was an active continuing programme of reform, intended to reorganise a wide range of institutions around certain political ideals.

The funds set up with Rockefeller capital were eventually merged to form one large foundation, which played a major role in shaping medical and university education, political reform, public health, social welfare, scientific research and agriculture throughout the world. Just as the Rockefellers had the savvy to hire good managers, skilful public relations people and astute lawyers for their business enterprise, so also the foundation recruited some remarkably farsighted and energetic scientific administrators. It has even been suggested that the notion that science can be managed, that goals can be set up and selective funding of research organised to realise them, was pioneered by Rockefeller Foundation officials in the 1930s, before the massive targeted programmes in applied science of the Second World War.

One such administrator was the exphysicist, Warren Weaver, who also became an expert on Lewis Carroll's *Alice in Wonderland*. In the 1930s, Weaver skilfully married the judgements of leading university scientists about promising fields of endeavour to a specific conception, held by dominant figures within the foundation, as to what counted as valuable, reliable and undersupported research. Weaver chose to use the enormous resources of Rockefeller philanthropy to "modernise" biology by insisting that researchers in various biological disciplines recast their research so as to make it more like physics and chemistry. Weaver felt that biologists were too content merely to classify organisms or to speculate in an undisciplined way about the mechanisms of evolution. Biology was either unambitious or undisciplined. It ought, he felt, to be placed on a more rigorous basis, tied in much more directly with experimentation and organised by carefully tested theories. Biology, as then practised, had no analytic power, no strength, no potential for reaching profound truths about the natural world. In Weaver's eyes, the physical sciences had just these virtues.

Consequently he was interested in importing the methods and technology of physics and chemistry into the life sciences. He was prepared to back people who could formulate a theoretical question, such as how genes copy themselves, and attack it in a concrete, practical fashion; for example he was attracted by the idea of thinking of the gene as a molecule, and then asking what kind of molecule a gene must be in order to function as a gene. Weaver was drawn to studies of large biological molecules, as a route to the understanding of function.

Because of his situation in the foundation he sought to cluster the projects he supported into a programme. He was keen on multifaceted research and was happy to allow projects to straddle the boundaries between disciplines. He strove to change biological research through an emphasis on lab equipment and research technology. Large sums of money went to people developing equipment, like the ultracentrifuge (which spins so fast that it grades molecules according to size) and the electron microscope (the most powerful microscope there is—sufficient, in fact, to show up viruses), for analysing biological materials in greater depth. Technology was a means of encouraging biologists to think differently and to ask different questions about organisms, questions that they might well have been unable to answer on their own.

He therefore put up money for fellowships abroad to allow people to pick up new skills, for conferences, for new laboratories and encouraged interdisciplinary collaborative work, against the fashions of the time. These things may not sound very fundamental, but with the resources at his disposal and the deliberate selectiveness of his support, Weaver's subtle pressures had an enormous effect. With his assistance, an avant-garde of "molecular biologists" (the name he gave them) set out on a long road to future glory. This was long before the era of big money for academic research. That was later forthcoming from the National Science Foundation and the National Institutes of Health in the United States, and bodies like the Medical Research Council in Britain.

This orchestration and restructuring of science had no specific industrial purpose. No particular payoff for the oil industry was intended. Nowadays things have changed somewhat. Exxon Research and Engineering—a separate institution, of course, from the Rockefeller Foundation—is funding a plant sciences lab at Cold Spring Harbor, the leading US molecular biology centre, to undertake industrially relevant research.

From the Wings to the Centre of the Stage

With the Second World War came a massive increase in funding for science and technology, not only to produce nuclear weapons, radar and new artillery, but also, medical innovations, such as the mass production of penicillin, antimalarial drugs and artificial blood plasma. Consequently, at the end of the war, people in government, industry and the professions realised what increased state expenditure on research could do. Medical research, including work in areas of biology that could only influence medicine in the longer term, began to receive far more cash than it had before. The money from the Rockefellers that had seeded new lines of investigation in the life sciences was gradually turned to research on seeds *per se,* which led to the high-yield rice and wheat of the 1960s (see chapter 5). Foundation money was dwarfed by government money rather than the other way around.

Some historians have described the massive increase in government support for research as a covert form of state subsidy to private medicine. On this view it represented a compromise between groups in society pressing for cheaper, more readily available health care and a reactionary medical profession trying to maximise their income by keeping a monopoly on medical services.

Gradually through the 1950s, research groups in molecular biology grew in a few centres around the world, several of them outside or only loosely linked to the university system: Cambridge, UK; Institut Pasteur, Paris; Cambridge, Massachusetts; Cold Spring Harbor, New York; Cal Tech and Stanford, California.

One school of molecular biologists busied themselves with analysing the structure of the molecular components of living things, such as haemoglobin, the oxygen-carrying pigment of red blood cells. That turned out to be a remarkably long struggle, with clear answers only appearing in the late 1950s, some twenty years after the work began. Others busied themselves with analysing one of the basic properties of genes: how they copy themselves and pass on the information necessary to produce a particular trait in the next generation.

One way of doing that was by trying to discover how viruses copy themselves. Viral infection comes about as these tiny entities penetrate living cells, hijack their internal machinery and turn it over to the production of more viruses. Since viruses are just tiny packages of protein wrapping up a few genes, this seemed a sufficiently simple system with which to explore the very general question of how like begets like.

Another group of molecular biologists focused on bacteria and began to breed them selectively to try to discover how particular properties, such as the ability to live on first one sugar then another, are controlled genetically. Bacteria have survived for longer than any other organism on this planet by being flexible, economical and amazingly opportunist. These tricks are clearly genetic. They are built into the genes of the bacterium. Finding out just how such abilities could be innate was to show more of what genes are and how they act in controlling living systems.

There are some common themes in all this which it would be as well to spell out. All these strands of research touched on fundamental or basic questions. Although they were funded by agencies interested in advancing medicine, their connection with health and disease was very abstract, general and long term. Secondly, in the 1950s all this work had yet to prove itself and gain widespread interest and attention. It was just a promising sideline. It took until the 1960s for the real power of molecular biology to become clear. Thirdly, the ability to move genes between different strains of bacteria (or more accurately to use natural mechanisms of gene exchange) was an important research technique, yet it was so specialised that nobody thought of it as "genetic engineering" in the full sense. The idea of having that kind of control over biology—especially human biology—seemed a very distant prospect. As we will see later on, that changed in the mid 1970s. Fourthly, molecular biology was not only separated from medicine, it also had very few industrial connections. Industrial microbiology was another world altogether and one that held no interest at all for molecular biologists. They were, frankly, a very self-assured community, and their pride in their academic achievements increased as the years rolled by.

Several major discoveries marked the journey from the margins of research to the centre of the biological stage. One of them came in 1944 when it was established that the material of which genes are made is deoxyribonucleic acid (or DNA for short). This discovery prompted a further set of questions, such as what the structure of DNA might be. If you know the substance of which genes are made, the obvious question must then be, how is that substance put together so that it will carry out the tasks of specifying hereditary traits and getting them transferred to the next generation? The answer to that emerged in 1953, when two cocky young scientists in Cambridge, England, Francis Crick, who was skiving from his Cambridge doctoral research, and James Watson, with his newly minted American Ph.D., put forward an elegant model based on the analysis of X-ray patterns that others had made and some reasoning about the possible configurations that subunits of DNA must adopt. This was the famous double helical model of DNA, now accepted as an accurate representation, although it is still occasionally queried and challenged.

At the time the double helix struck only a limited group of scientists as really interesting. Many biologists schooled in the traditional disciplines of botany, zoology, physiology and classical genetics went their own sweet way, to the dismay and contempt of molecular biology zealots, who could see all kinds of problems and new approaches to be explored at the molecular level. In Manchester, for a while in the 1950s, the zoology department actually banned teaching about DNA as an irrelevance.

As the zealots saw it, biology could at last be placed on a firm physico-chemical foundation. Romanticism about living nature—about plants and animals—would henceforth give way to hard science. They were to create a cult of elegant reasoning from a few carefully planned experiments which sought to lay bare a new set of abstractions about very general biological mechanisms. This was a perspective rooted in physics and genetics, both highly theoretical disciplines. The zealots made no secret of the fact that they felt their gaze went further and deeper than that of other disciplines. They had no qualms about calling on scientists to rethink their research in terms drawn from molecular biology, with its emphasis on structure and information. It was a posture of breathtaking arrogance and provocativeness, yet it drew upon the knowledge that achievements like the postulation of a double-helical structure to DNA were scientific feats of a high order. These really did indicate the fruitfulness of thinking in molecular terms. It was not a boast without substance, but it was still a boast.

History tells us that Watson and Crick proposed the double helical model of DNA in 1953. The Nobel prize awards for this work in 1962 divided the credit three ways, with the recognition of the contributions of Maurice Wilkins at King's College, London. In recent years various writers have pointed out just how much all of them owed to the work of a colleague of Wilkins, Rosalind Franklin, who remained unaware, up to her early death in 1958, of the extent of this debt and the means by which her data were passed on. Moreover, Watson's and

Crick's final inferential steps were possible only because of the accumulated labours of perhaps thirty or forty other scientists, including a few who gave them crucial information that put them back on the right track.

James Watson's retrospective account, *The Double Helix,* was very candid indeed about the competitiveness, arrogance and delight in scoring off senior colleagues which characterised the path to what they only half-facetiously called "cracking the secret of life." His candour caused great embarrassment: not least to Crick, who persuaded Harvard University not to publish the book, and who threatened to sue the eventual publishers for libel. But many observers have pointed out that, in spite of its many indiscretions, Watson's account was salutary in showing just how far from the ideals of scientific cooperativeness and community the reality of pursuing a scientific career is. Most retrospective accounts are elegant and noble; Watson told it like it was in the trenches. The book was also, inadvertently, very self-revealing about the distortions and absences in the character of people who are out to "make it" through science.

The double helix immediately suggested a new set of questions that could be explored to deepen our knowledge of how genes act in passing on traits from generation to generation. If genes were part of a complex, but highly regular, molecule composed of two strands twisted around each other, then gene replication—the copying of a set of instructions for the next generation—could be thought of as the separation of the two parts of the helix and the reconstitution of two separate double helixes, each built on one of the separated stands.

To think about DNA we have to represent it to ourselves. That is difficult because molecules, being so minute, have different characteristics from much larger objects like chairs and tables. They don't have definite boundaries, yet we have to draw them as if they had. Also DNA is an astonishingly intricate molecule. You can easily get lost in the detail. So you have to be schematic. The figure on the next page is one compromise between accuracy and ease of comprehension. The essential feature to pick out is the two helical strands, here called Ribbon I and Ribbon II. These separate when genes are copied to make a new generation. Making that clear was the power of the Watson-Crick model.

But this left the intriguing question of how some structural feature of the DNA molecule was able to represent or encode a particular hereditary trait. Now the notion of a genetic code antedates the double helix by nearly twenty years. Furthermore, the perception that the

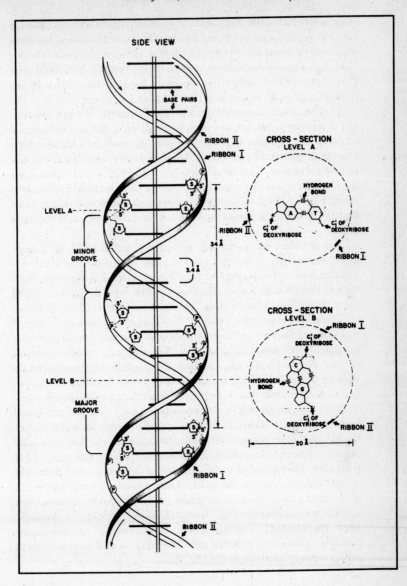

structure of the newly discovered double helix could serve as the basis for such a code came not from a biologist in the first instance, but from an eccentric cosmologist, George Gamow. Being a complete outsider to biology actually helped him to focus on the fact that one could view that apparently disordered sequence of bases at the centre of the double helix as an immensely long number.

In other words one could think of DNA primarily, not as a constituent of a living organism, not even as a molecule, but in a completely abstract way, as a certain kind of number. It would be a number long enough to represent the vast range of characteristics that make any individual unique. Gamow got in touch with Watson and Crick about his ideas for a genetic code, and together they refined it with some success. By building on Gamow's initial insight Crick was able to show that the genetic code must have certain features and could not have others.

The problem was eventually solved in the mid 1960s by a more direct experimental method. Success came with a mad scramble for priority, leading to a further round of Nobel prizes. Furthermore, feats such as this produced a whole genre of press reports sensationalising the work with headlines such as "The Secret of Life Has Been Broken."

These successes with some of the basic theoretical problems of molecular biology came around the same time as some more concrete, but equally important, achievements. For example, the quest to work out the three-dimensional structure of a protein, first sponsored by Weaver in the late 1930s, was realised in the early 1960s. In France, two molecular biologists, François Jacob and Jacques Monod, showed, in 1961, how the production of proteins within a bacterial cell could be controlled genetically. They postulated the existence of a sophisticated switching system that served to regulate whether particular substances were to be made in the cell, under what conditions, and at what rate. They pointed to self-regulating or "cybernetic" networks in bacteria, centred on the genes. Cells were thus programmed to adapt their activity to environmental conditions and to regulate their activity with self-correcting "feedback loops." Finally, in the early 1960s, the scientists working with viruses came to the first clear understanding of how they infected cells like bacteria and caused them to produce many more copies of the virus.

Competition, Power and Patronage

By the late 1960s, molecular biology had achieved a powerful position within the life sciences. Its practitoners began to pick up a surprisingly large proportion of the Nobel prizes in physiology or medicine, and this conferred additional status on the molecular biological approach. Principally in the United States, money began to pour into the field, not least because of a connection with cancer research. The powers of patronage of leading molecular biologists grew accordingly. Having developed from the marginal pursuit of a few far-flung pioneers braving the disapproval of their peers, molecular biology became the site of some of the most intense competition in science and acquired a reputation as a field in biology where the real excitement lay and where the most daunting problems would continue to be found. It attracted many young scientists, not all of them with a biological training, who were in search of fame, if not fortune. Just as some of the early pioneering molecular biologists, like Crick, had been physicists, with sufficient self-assurance to attempt to instruct biologists in new ways of thinking, so too some of the later recruits were undeterred by the lack of a basic training in biology. They believed that they could gain immediate access to a higher level of problems.

Already, in the early 1960s some of the friendly camaraderie was giving way to a hardened determination to reach answers first. By the end of that decade a highly competitive ladder of achievement had been built, so that research students who wished to stay in the field had to prove their competence again and again, often in different labs, in order to progress towards an established post, with their own research subordinates and membership of a highly self-regarding élite with plenty of disdain for lesser mortals. The men who stirred up this kind of competition were some of the early pioneers, whose lobbying for resources had brought them large, well-equipped labs, maintained by grants from government agencies, and who year after year sifted through the available talent, scanned the research fronts for new ideas and drove their research teams hard to keep the turnover of grants, results and trainees going. One such research director, James Watson, sometime Harvard professor and now head of the prestigious laboratories of Cold Spring Harbor in New York State, has compared the regime to that in American pro football where young athletes are pushed to peaks of achievement to keep a team's standing in the major leagues.

Such intense effort has brought impressive results. This tunnelling into nature at the level of molecular organisation has yielded an enormous amount of information about some of the basic processes of life, such as heredity, metabolism, infection, cellular pathology and growth. The molecular approach has spread throughout the life sciences, with different effects in different fields, transforming methods, attitudes, models and theories as the wave has advanced. In some areas like embryology it has not been very useful. No one knows how an eye becomes an eye. In others, like immunology, it has proved extremely productive, as we shall see in chapter 4. Molecular biology has led to a change in the way all kinds of problems are formulated and attacked.

The double helix is now in elementary textbooks (though it is usually represented inaccurately). From an early stage there are cues and incentives to think molecular and to consider organisms as "self-assembling, self-maintaining, self-reproducing information-processing machines." The idea that the genes embody *encoded information*, which is first *transcribed* into a chemical intermediate, *messenger RNA*, and then read *codon* by *codon* and *translated* into a *sequence* of amino acids to form a protein molecule is basic to the conceptual framework of modern biology. Genes are *switched on* and *off;* messenger RNA is *processed* and *edited;* gene sequences contain *signals* to the cellular *machinery;* and information is *read out* from the DNA. What this means in detail is the substance of the next chapter; it is the metaphorical language that is speaking to us here. The metaphors are informational; the idioms, without which one can scarcely think or talk biology these days, are drawn from computing, cryptography and cybernetics; the hardware and software of information technology is projected on to the cell. All this is a powerful kind of shorthand for thinking about what living things do. These notions have become basic to the vocabulary and analytic perspective of most, but not all, biologists.

By the end of the 1960s the first phase of expansion of molecular biology was over. Some of the key issues, like the nature of the genetic code, had been sorted out, which made the selection of the next set of comparable problems rather difficult. What questions could be as important? The early 1970s were also a time of some anxiety in science. Research budgets were starting to fall in real terms for the first time since the war. Money was available in the field of cancer research, but all the approaches to that problem looked formidably difficult.

Then a series of technical discoveries came together that have completely altered the landscape of possibility in molecular biology. What became practicable, for the first time, was the controlled manipulation of pieces of genetic material. Researchers could snip out sections of DNA from one organism and transfer them to another. One could cut genes out of one cell and splice them into another. The

technical term for this activity is "recombinant DNA research" because what is involved is the controlled "recombination" of sections
of DNA, the hereditary material. (We shall go more deeply into the
technical detail in the next chapter.) The implications of this for many
areas of research were staggering. For example one could now think
about splicing new bits of DNA into tumour viruses, or taking them
out with precision, to see how that modified their impact on the cells
they infected. But as the word spread of the technical possibilities, a
group of research directors, the laboratory heads that controlled the
profession, decided, with some urging from the grass roots, that "progress" should be halted for a while, in order to see just what unintended
consequences these tricks might have. When that moratorium was
lifted in 1976, a new era of genetic engineering began and with it, in
all likelihood, a new industrial and scientific revolution.

The Limits of Public Participation in Science Revealed

The Moratorium in Context

It would be sad if history does not underline the remarkable fact that
unlike the computer, nuclear reactors, rocketry and satellites, the
technology of genetic engineering was stopped in its tracks right at its
point of inception. From 1974 to 1976, there was a worldwide deferral
of work in this area. Why should this be missing from future histories?
Because some of the guys involved are now doing their damnedest to
disavow their role in setting up a moratorium for those two years.
They feel that this action was a youthful folly, even a terrible error of
judgement. As the inimitable James Watson put it, "I was a jackass."
As they see it, the events set in motion by the moratorium were a
parody of serious debate, a string of public discussions of soon-discred-
ited, uninteresting, jettisoned hypotheses. These centred on the possibility of turning relatively harmless micro-organisms into extremely
dangerous ones by inserting new genes into them in the form of
recombinant DNA molecules. Month after month, in city after city,
much the same group of critics and supporters of this work foregath-
ered to rehearse yet again the idea that splicing new genes into recipient micro-organisms could have unexpected medical, environmental
or biological consequences.

Watson now clearly regards this phase as over, which is probably
correct. He and another former combatant against sensationalist de-

bate, John Tooze, have just issued a history of this episode, as if to mark its end. The sense of relief that time need no longer be spent trying to tackle complex social and political issues is almost palpable in their book. They seem to have come to experience the public debates as a kind of political hydra, with the capacity to grow new heads when one was removed by a particularly concerted rhetorical swipe. In the late 1970s Watson waxed very indignant about what he saw as the craziness of being worried about the dangers of recombinant DNA. How could people focus on such a trivial matter, which molecular biologists now *know* to be safe?

The problem, of course, was one of legitimacy and credibility, and Watson may not have realised it. By the early 1970s scientists could not count on reflex public acceptance of their pronouncements, particularly on issues concerned with heredity, disease, reproduction and the environment. The basis of trust between experts and the general public had been transformed by a decade of mounting concern about the implications of biomedical research and the related evidence of unintended, unfortunate consequences of technological innovation. Consensus about acceptable risks and moral justification for particular programmes of research had to be nurtured and negotiated far more carefully than had once been the case.

At some level scientists were aware of this. Yet negotiating such issues takes time. Questions about moral values, implicit assumptions and goals and the possible consequences of research cannot be dealt with quickly or straightforwardly. They are a skein of issues that gradually unravels as they are considered, so that the boundaries of scientific competence and authority become blurred and ambiguous—and open to challenge.

Accordingly, the moratorium and the ensuing debates about recombinant DNA research have to be seen as an experiment in discussion with the public that looked to scientists as if it would hurtle out of control. As public interest in the issues built up, scientists' alarm at where that interest could lead also increased. The most pressing question for the majority became how can we get back to the status quo, in which public interest in science was admiring and quiescent, and scientists' competence to manage the social implications of science was relatively unquestioned. What started out as a noble gesture of visible reflection about risks threatened to become an exercise in power-sharing and that was something that biologists were determined to resist. The question of how the growing powers of molecular genetics would be used, and whether society could cope with these

Promethean abilities, had surfaced regularly in the 1960s. Much of the discussion, represented by titles like *The Biological Time Bomb, The Prometheus Project* or *Brave New Baby,* was doom-laden and irritatingly vague; mysterious emanations from unnamed laboratories made by unidentified individuals were apparently going to sweep over society. Few people set out to ask who controls science and who in fact controls its applications. Nor did any of the authors of the sensational tracts consider how specific groups in society, at risk from science, could formulate a political response to protect their position and even nullify the threat facing them.

An honourable exception to this generalisation came in 1969. Three Harvard biologists made the announcement of their isolation of a bacterial gene the occasion for a press conference. They pointed out that the control of science in the USA lies in the hands of a tiny élite. Its members constantly coordinate the planning of science with representatives of major corporations and members of the military and political establishments. By this means, the Harvard biologists claimed, science is systematically abused through its application to warfare and the pursuit of profit. One of these people, Jonathan Beckwith, later went on to donate money from a prize honouring this work to the Black Panther Party, to show the strength of his feelings that science, far from helping the disadvantaged, was systematically being used for their continuing oppression.

The Harvard press conference received major media coverage, but the point that society at large has no real control over the direction of science got lost, as commentators wilfully misinterpreted their statement as yet another outburst of nonspecific concern about genetic engineering. They also mislaid the accusation that the scientific establishment, far from trying to influence the general pattern of application of science, colludes in the abuse of research for inhuman and exploitative ends. Research élites assist social domination through science, rather than acting as tribunes of the people. The implication was that the power of leading researchers could be channelled only in certain directions. They are a captive élite, and they misrepresent (to a trusting public) their degree of autonomy and their ability to handle social problems occasioned by science.

This view was thrown sharply into relief by an opposing one put forward by leading scientists at an important conference in 1970, on "The Social Impact of Modern Biology." Under the auspices of the newly formed British Society for Social Responsibility in Science, a group of distinguished scientists, including several Nobel prizewinners, addressed themselves to such questions as the prospect of human genetic engineering, *in vitro* ("test tube") fertilisation, the marketing of pharmaceuticals, chemical and biological warfare and the use of scientific theories to legitimate political ideologies. Whilst Beckwith and others called for major structural change in society to erode the power of those presently controlling science, the Nobelists with a common voice argued that the management of science and its social problems was best left with leading scientists. The scientific mandarinate claimed the right and the responsibility to deal with these things as its members saw fit. Those without impeccable scientific credentials had a right to hear the results of such deliberations about the likely impact of science on society, but not to participate in them or to set the terms of such appraisal.

Perhaps the most audacious deployment of this idea came from the French Nobel prizewinner Jacques Monod. He was a charismatic figure, with a distinguished record as a Resistance fighter, and was the author of a book on the existential and political implications of molecular biology, *Chance and Necessity,* which outsold Erich Segal's *Love Story* when it appeared in France. In 1970 Monod was just turning from a brilliant research career to assume the directorship of the Institut Pasteur in Paris. His plans for the more intensive commercialisation of work done there and the sale of the buildings in Montparnasse, where Louis Pasteur had worked, created uproar. Monod was then a man of tremendous faith in his own abilities, not least to manage the future on behalf of society. At the "Social Impact of Modern Biology" conference he mentioned certain potentially worrying developments on the horizon which would revolutionise genetics, but stated that he preferred not to speak about them in public. It is almost certain that he was talking about the first intimations of recombinant DNA research. Apparently the statement of his private concern with these possibilities was supposed to allay any public anxiety. It was as if a member of the mandarinate had spoken, and that was sufficient.

But while Nobel laureates felt no embarrassment about excluding the public from a consideration of the social impact of molecular biology, they were happy to point out other people, involved in the commercialisation of research, who should come in for some criticism. Another speaker, A. J. Hale, of the drug company G. D. Searle, was roundly abused by leading scientists for his association with such dubious commerce. Ironically it was Jonathan Beckwith who, while not exactly defending Hale, pointed to the hypocrisy in this response. Searle now has a large research group and a production plant at High

Wycombe devoted to recombinant DNA research, in which Hale is closely involved. Plans for such a laboratory were already being discussed in 1970. Whether any of his distinguished critics at that event now act as consultants to Searle, I don't know. I doubt it, but it is not impossible, as many scientists do advise commercial concerns and own shares in them.

The 1970 "BSSRS" conference was a heady occasion—blessed by a large handful of Nobel prizewinners. The clash of views was fascinating to watch and very educative. The meeting marked a watershed in the re-emergence of a social movement concerned with the control of science and signalled the imminent parting of the ways between the scientific establishment and the more avowedly socialist among the scientists. The same questions of power, responsibility, expertise and accountability came up again and again in the 1970s, in the debate over recombinant DNA research. Gradually, however, the establishment position has hardened, as the political and economic costs of engaging in such precarious negotiations have risen. If such conferences and meetings take place, they do so, as we shall see, very much on terms set and controlled by the mandarinate or the corporate sponsors. The stakes are now too high, it seems, to let participation go very far. But in the mid 1970s it really seemed for a while as if the ways that the social implications of biomedical research are handled were about to change.

Signals of Concern

In 1973, various groups in the United States semiformally announced at a scientific conference their plans to splice together genes from different organisms. Just how this was done we shall consider in the next chapter. This novel idea immediately excited concern. After all nobody could say for certain what the biological results would be. It seemed just possible that the hybrid viruses envisaged in one line of experiments could emerge with the ability to cause cancer in hitherto unaffected organisms. Equally it was not quite beyond belief that, unintentionally, a prodigious and drug-resistant dose of nastiness could be built into laboratory bacteria, which could become the agents of a pandemic. The first discussions of these unlikely, but disturbing, ideas led to the formation of an international cell of scientific strategists, who were asked by the National Science Foundation in the United States to come up with a plan.

In 1974, this group, led by Paul Berg, a molecular biologist from

Stanford, California, and the man whose plans to splice bits of virus together had sparked off the concern, proposed a self-denying ordinance for would-be genetic engineers. They recommended holding off from certain kinds of experiment, until the possible risks could be specified with greater precision. This call for a moratorium, conceived as a delay, and not as a prelude to total renunciation, was immediately accepted and backed up with various bureaucratic incentives to make everyone conform. In fact one or two wild spirits pressed on regardless. But in general publication would have meant damnation, not least for having the gall to try and steal a march on the competition.

Some writers argue that the moratorium sprang from a collective desire to be seen in public to behave with propriety, against the dismal background of the war in Indochina and the Nixon presidency, which many Americans felt had polluted public life. Others have said that the signatories of the Berg letter, in which the moratorium was suggested, were less noble and merely wanted a place in the history books. No matter; either way it was a remarkable gesture. It set an important precedent and it attracted a great deal of comment. Because it is a historic document it is reproduced in full here. Notice that it does not call for public participation in science, although that was its immediate effect. Gradually, the elements of a full-scale "public debate" were activated and the discussions of what to do next intensified.

Potential Biohazards of Recombinant DNA Molecules

Recent advances in techniques for the isolation and rejoining of segments of DNA now permit construction of biologically active recombinant DNA molecules *in vitro*. For example, DNA restriction endonucleases, which generate DNA fragments containing cohesive ends especially suitable for rejoining, have been used to create new types of biologically functional bacterial plasmids carrying antibiotic resistance markers and to link *Xenopus laevis* ribosomal DNA to DNA from a bacterial plasmid. This latter recombinant plasmid has been shown to replicate stably in *Escherichia coli* where it synthesizes RNA that is complementary to *X. laevis* ribosomal DNA. Similarly, segments of *Drosophila* chromosomal DNA have been incorporated into both plasmid and bacteriophage DNAs to yield hybrid molecules that can infect and replicate in *E. coli*.

Several groups of scientists are now planning to use this technology to create recombinant DNAs from a variety of other viral, animal, and bacterial sources. Although such experiments are likely to facilitate the solution of important theoretical and practical biological problems, they would also result in the creation of novel types of infectious DNA elements whose biological properties cannot be completely predicted in advance.

There is serious concern that some of these artificial recombinant DNA molecules could prove biologically hazardous. One potential hazard in current experiments derives from the need to use a bacterium like *E. coli* to clone the recombinant DNA molecules and to amplify their number. Strains of *E. coli* commonly reside in the human intestinal tract, and they are capable of exchanging genetic information with other types of bacteria, some of which are pathogenic to man. Thus, new DNA elements introduced into *E. coli* might possibly become widely disseminated among human, bacterial, plant, or animal populations with unpredictable effects.

Concern for these emerging capabilities was raised by scientists attending the 1973 Gordon Research Conference on Nucleic Acids, who requested that the National Academy of Sciences give consideration to these matters. The undersigned members of a committee, acting on behalf of and with the endorsement of the Assembly of Life Sciences of the National Research Council on this matter, propose the following recommendations.

First, and most important, that until the potential hazards of such recombinant DNA molecules have been better evaluated or until adequate methods are developed for preventing their spread, scientists throughout the world join with the members of this committee in voluntarily deferring the following types of experiments.

• *Type 1:* Construction of new, autonomously replicating bacterial plasmids that might result in the introduction of genetic determinants for antibiotic resistance or bacterial toxin formation into bacterial strains that do not at present carry such determinants; or construction of new bacterial plasmids containing combinations of resistance to clinically useful antibiotics unless plasmids containing such combinations of antibiotic resistance determinants already exist in nature.

• *Type 2:* Linkage of all or segments of the DNAs from oncogenic [cancer-inducing] or other animal viruses to autonomously replicating DNA elements such as bacterial plasmids or other viral DNAs. Such recombinant DNA molecules might be more easily disseminated to bacterial populations in humans and other species, and thus possibly increase the incidence of cancer or other diseases.

Second, plans to link fragments of animal DNAs to bacterial plasmid DNA or bacteriophage DNA should be carefully weighed in light of the fact that many types of animal cell DNAs contain sequences common to RNA tumour viruses. Since joining of any foreign DNA to a DNA replication system creates new recombinant DNA molecules whose biological properties cannot be predicted with certainty, such experiments should not be undertaken lightly.

Third, the director of the National Institutes of Health is requested to give immediate consideration to establishing an advisory committee charged with *(i)* overseeing an experimental program to evaluate the potential biological and ecological hazards of the above types of recombinant DNA molecules; *(ii)* developing procedures which will minimise the spread of such molecules within human and other populations; and *(iii)* devising guidelines to be followed by investigators working with potentially hazardous recombinant DNA molecules.

Fourth, an international meeting of involved scientists from all over the world should be convened early in the coming year to review scientific progress in this area and to further discuss appropriate ways to deal with the potential biohazards of recombinant DNA molecules.

The above recommendations are made with the realisation *(i)* that our concern is based on judgements of potential rather than demonstrated risk since there are few available experimental data on the hazards of such DNA molecules and *(ii)* that adherence to our major recommendations will entail postponement or possibly abandonment of certain types of scientifically worthwhile experiments. Moreover, we are aware of many theoretical and practical difficulties involved in evaluating the human hazards of such recombinant DNA molecules. Nonetheless, our concern for the possible unfortunate consequences of indiscriminate application of these techniques motivates us to urge all scientists working in this area to join us in agreeing not to initiate experiments of types 1 and 2 above until attempts have been made to evaluate the hazards and some resolution of the outstanding questions has been achieved.

Paul Berg, *Chairman*
David Baltimore
Herbert W. Boyer
Stanley N. Cohen
Ronald W. Davis
David S. Hogness
Daniel Nathans
Richard Roblin
James D. Watson
Sherman Weissman
Norton D. Zinder

Committee on Recombinant DNA
Molecules, Assembly of Life Sciences,
National Research Council,
National Academy of Sciences,
Washington, DC 20418

Characteristically, in Britain a working group of distinguished scientists was set to meditate on the implications of the moratorium, shepherded by Lord Ashby, a distinguished biologist, and sometime Master of Clare College, Cambridge. So tightly is the British scientific establishment structured that, when their report appeared in February 1975, one man, perched on many official committees, was asked to reply in two different guises to the document that he had helped to draft. The Ashby report planted in the international scientific consciousness the ideas of limiting experiments to enfeebled strains of bacteria and of a national advisory group which would coach scientists to levels of safe laboratory practice that they otherwise might not have bothered to attain. These eminently reasonable suggestions were energetically disseminated at the Asilomar conference in California in March 1975.

This was the convocation proposed in the letter from Berg and his colleagues. It was to formulate an informed and coherent position on the risks of genetic manipulation. To be invited was a sign that one belonged to an exclusive scientific fraternity. Among biologists the meeting has attained a kind of mythic quality. Some people speak of it the way, in other circles, it is said, "I was in Cable Street when they stopped Mosley" or "I was at Vatican One." I suspect that it was like many other international scientific conferences, with a lot of trade going on informally, and some sly feeling-out of work in competitors' labs. It is equally probable that the formal sessions were dominated by a restricted group of voluble and ever more confident speakers developing their roles of clown, elder statesman, Young Turk, sceptic and gadfly, and that a hard-pressed female secretariat had to get pampered and rather helpless males on to their next international flight. It was an exclusive meeting, though some members of the press were allowed in. The whole thing was tape-recorded for posterity. The best account of it appeared shortly afterwards in *Rolling Stone* magazine, which showed just what the so-called New Journalism could do for the reporting of science.

Michael Rogers' article featured some of the power ploys going on in the conference. He described a few of the more flamboyant gestures and the weariness of the organisers, facing the prospect that their efforts might come to nought if the meeting ended without a consensus. He showed how totally ignorant the US scientists were of their legal liabilities as directors of laboratories where dangerous materials were handled. Rogers' irreverence and refreshing refusal to be awed by the symbols of authority allowed him to get closer to the political realities of science than more docile journalists ever do. But he made the banalities of the conference into something special, and for that even the participants loved it too. If only flummery, guile, quirkiness and self-importance in academe could be exposed more often.

The legacy of Asilomar was the idea of guidelines that could regulate recombinant research, by specifying what kinds of safety precaution should be taken for a given experiment. At this stage the goal was to ensure that work with recombinant micro-organisms on the laboratory bench should be no more dangerous than any other experiment in microbiology. It is worth remembering that although no one made much of this in the mid 1970s, research on genetic manipulation had already begun in industry. But it was the academics, unused to industrial safety practices, who were so eager to press on and to define their own safety criteria. The main issue was, How do we know that playing around with genes in this way is safe? The problem for the impatient scientists was that increasing numbers of people answered that question by saying, "We are not convinced that it is safe," or even from the more confident souls, "We know that you cannot know that these experiments are safe, whatever you might say at this stage."

Public Pressures for Regulation

Indeed on the day that a set of guidelines for US researchers were issued by the National Institutes of Health, a full-scale confrontation between academic scientists at Harvard University and representatives of the local Cambridge community took place. On the one side were Harvard molecular biologists, aching with the desire to establish a new laboratory for the latest gene-splicing work. On the other was the mayor of Cambridge, Alfred Vellucci, and a coalition of concerned citizens. Vellucci, an unashamed populist, appeared to be perfectly happy to exploit the issue, no matter how absurd scientifically, to focus attention on what he saw as just another episode in the exploitation of his city by an immensely wealthy and disdainful university.

Under the public health regulations, proposals for new laboratories could be discussed by the city council. The request to build a new P3 lab (so called because of the strict standards of physical containment embodied in its construction) from Harvard biologist Mark Ptashne was considered at a highly charged public meeting in July 1976. Amongst other things it emerged that the building within which the lab was to be created was infested by radioactive ants that had defied eradication for some time. If this was so, could one really talk, the sceptics asked, of high security containment of bacteria, which are the kind of thing with which the ants could cover themselves and wander off?

In the event, a nonexpert citizens' committee was set up to examine the problems, and while it did so a mini-moratorium was imposed on work at Harvard. Its report was a model of common sense and the new laboratory duly went ahead, but not before some scientists had flown to campuses in less demanding communities. Throughout the entire saga of safety regulation this has been a recurrent threat from scientists and corporations, particularly from the hotheads: "If you won't let us do it here, we'll just move to where the natives are friendlier." At one conference in 1979, Charles Weissman, a researcher from Zurich and now involved with the biotechnology company Biogen, showed a cartoon in a technical presentation (a sure sign of a secure reputation) of well-known scientists at an airport check-in desk, each with a thermos containing his research bacteria, each going off somewhere different. His audience chuckled delightedly at the flight from persecution to the Promised Lab.

By the autumn of 1976, two sets of guidelines for recombinant DNA research existed, one created in the US, encyclopaedic, complex and rigid, the other by another UK government committee, which was flexible, pragmatic and modelled on case-law rather than on statute. Throughout the developed world, governments opted for one model or another or some mixture of the two. Committees were set up to scrutinise and refine the regulations. The American one meets in public, publishes its minutes, allows nonmembers a period for comment in its meetings, and no curtain is drawn over the periodic clashes on the committee. It is now called the Recombinant DNA Advisory Committee (RAC). Scientists speak of "being on the RAC" in an unmistakable allusion to medieval torture.

The British committee, the Genetic Manipulation Advisory Group (GMAG), on the other hand, meets in official seclusion and sworn secrecy, and publishes no minutes, although several annual reports have appeared. New members sign the Official Secrets Act and receive powerful indoctrination in the codes of discretion expected of visitors to the British governmental machine. But unlike the RAC, at least in its original constitution, GMAG has trade union and "public interest" representatives on it.

In the United States, trade unions representing nonmanagerial, technical personnel, including university technicians, have often been harassed or ignored, so that organised technical labour is no force to be reckoned with. In the UK, on the other hand, white-collar trade unions have recruited very successfully since the early 1960s. Organisations like the Association of Scientific, Technical and Managerial Staffs (ASTMS) exist as loosely bound groups of technical workers, held together by a professional full-time staff that carefully nurtures and services the economic militancy of the membership. Under the guiding hand of its secretary-general, Clive Jenkins (once the bane of the British upper classes, but now a more mellow corporatist), ASTMS has won recognition of its legitimate interest in health and safety at work, even from obdurate and conservative employers like British universities and government laboratories.

Occupational Hazards of Lab Science

In the mid 1970s, the occupational hazards of working with dangerous micro-organisms were a matter of concern to lab workers in hospitals, research establishments and universities. In 1973, several people died of smallpox at the London School of Hygiene and Tropical Medicine. In 1978, a technician contracted smallpox from a poorly run virology lab at Birmingham University, where local recognition of ASTMS had been resisted, and where working practices were supposedly checked by a government committee of experts. This body, the Dangerous Pathogens Advisory Group (DPAG), had clearly taken on trust the assurances of a close colleague that his lab was safe, when in fact he was working frantically, in conditions that his peers would have found unacceptable, in order to finish some work before the money ran out. The accident at Birmingham discredited the model of advisory safety committees made up only of leading researchers. It also led to reform of DPAG itself in which trade union representatives are now included.

Even so, these changes had to be fought for. They did not follow straightforwardly from a simple realisation that the regulatory system did not work, and many of the wrangles and attempts to keep things as they were took place out of public view. The Department of Health and Social Security commissioned an investigation into working conditions in the virology unit by Professor Shooter, himself a DPAG member at one point. His report revealed a number of irregularities in the lab, but it only reached the public domain through leaked publication in the ASTMS periodical, *Medical World.* There was a threat of prosecution under the Official Secrets Act, but it never reached the courts. ASTMS then brought an action against Birmingham University on behalf of the woman who died there. In that case, it proved impossible to pin down responsibility for safety standards in the virology unit (its director having committed suicide), since Mrs. Parker's infection by smallpox could not be definitely linked to the work in the lab on the floor below her office.

The late 1960s and early 1970s saw growing scepticism amongst the lay public and younger more radical scientists in Britain about the ability of senior scientists to control the hazards that work in their laboratories was creating. In the same period laboratory workers, facing hazards at work from infectious organisms, chemical solvents, radiation and radioactive materials, became increasingly impatient with the claims of their distinguished bosses, to be seriously concerned with or totally competent to deal with lab safety. The Birmingham incident exposed starkly and tragically how little such directorial pronouncements of concern are sometimes worth. Coming as it did some way through a trade union campaign for more involvement in safety regulation, it served to reinforce what many lab workers had claimed for a long while; that ambitious, highly specialised research scientists cannot be allowed to define what are acceptable risks for their subordinates to run.

With battles of this kind going on and a significant section of its membership working in technical posts in universities, government labs and industry, ASTMS organised a conference in London on genetic engineering in October 1978. This has turned out to be the only major public discussion of recombinant DNA research in Britain.

Rolling Back Public Participation in Science

As recombinant DNA research began to accelerate in the late 1970s with the lifting of the moratorium, more and more interest came to be focused on the questions of possible biological hazards, though the vast bulk of public discussion of this issue occurred in the United States. Principally, this happened in cities containing one or more major universities, or at specially constituted state or federal governmental hearings. When the possibility of specific legislation which would have defined standards of practice expected of recombinant DNA researchers was canvassed in the US, a powerful lobby acted to roll back the congressional forces sympathetic to this idea. Just how much money was spent on this is not clear, but a great deal of time, energy, and professionally directed agitation was devoted in 1977 and 1978 to discrediting the proposal that specific statutes were necessary to minimise the hazards.

When this issue came up in the Federal Republic of Germany in 1980, a whole gallery of people with markedly different views about genetic engineering was assembled in Bonn. Each day they were ferried by bus, through the machine-gun emplacements that surrounded the Federal Ministry of Research and Technology, into a conference hall, where parliamentarians listened politely to the conflicting opinions. One could scarcely call this a public debate, although a fat volume was eventually issued describing the proceedings. There is no law specifically regulating recombinant DNA research in West Germany.

One effect of the scientists' burgeoning criticism of measures being taken to control the possible risks of genetic manipulation was to stimulate leading molecular biologists to set up an international committee to emphasise the value of genetic manipulation and the need for minimal regulation. This committee, known as COGENE, played a leading role in feeding data on the risks of research into the public debate and in orchestrating subsequent users of that information.

COGENE became, in effect, a pressure group for minimal regulation of this research, and its members exploited all their connections in governments around the world to get the message across. All in all, it was COGENE that built the consensus that recombinant DNA research is safe. This is not to impugn its members' scientific integrity in any way, but it does present them as a group of energetic, well-supported scientists and administrators acting in concert to secure a political objective, in a way not at all obvious to the general public.

Another response to public and professional anxiety about the potential hazards of recombinant DNA research was an attempt by some scientists to quantify the risks involved. Surely, they argued, if we could put a number on them it would help us to see how seriously they should be taken? The analytic techniques involved were already in use in the chemical and nuclear industries.

Essentially the task involved is one of trying to specify every element in a complex system and to figure out how likely it is to go wrong. You also have to work out all possible combinations of failures and problems. But as the nuclear power accident at Three Mile Island showed in a spectacular way, complex technological systems simply defy such analysis. Bacteria are every bit as complicated. Despite that, some scientists seized upon the idea of "quantitative risk assessment" and used it to show that people were far more likely to die of food poisoning than from possible mishaps at their local genetic engineering lab. Even if that could be shown to be true—and the evidence on which *any* argument in this field can be based is meagre—it does not follow that the public should give up their interest in how the safety of genetic manipulation is assessed and monitored. After all, the chances of catching smallpox these days are small, but the possibility is still a matter of public concern.

In 1979, four years after the Asilomar meeting, the COGENE people organised a private conference at Wye College, the agricultural outstation of London University in Kent. Just as at the earlier meeting, the whole of the formal proceedings were tape-recorded including some stentorian snuffling into the microphone, burps, sick jokes, *faux pas* and insults. (The published record has been cleaned up.) The press were intially excluded, then three delegated reporters were given permission to come, only to find themselves under attack when they spoke.

Much of the conference was given over to loving discussion of the latest technical achievements, but some time was devoted to risk assessment. Just how long could enfeebled laboratory micro-organisms stay alive in the human intestines if one swallowed them, say, by having a quick coffee in a lab where someone had inadvertently generated a mist of bacteria? Could they survive in sewage or down the drains if some naughty person swilled them down the sink?

The object of this and more arcane matters was to convince the participants that the existing regulations for handling such bugs were far too strict and that the meeting should agree to the publication of a strong statement calling for the guidelines to be considerably relaxed. In the event that did not happen. One or two influential sceptics dug in their heels and declined to support the composition of such a public statement. That was enough to weaken the take-home message, and the organisers were visibly incensed by this setback to their carefully laid plans.

Throughout the meeting there was a certain weary impatience with the view that the safety of all recombinant DNA research had not yet been definitely established. The conference had the air of a private get-together for scientific heavyweights worn down by endless rounds against a mindless opposition. When the US civil servant who serviced the RAC (the Advisory Committee) gave a talk on behalf of the director of the National Institutes of Health (kept in the United States on the orders of President Carter, because of events at Three Mile Island), he punctuated it with slides of the critics of recombinant DNA research caught in undignified or comical poses. The audience loved it, as they were supposed to. It was a most extraordinary performance by a supposedly neutral public official. It was perhaps matched by the new chairperson of the RAC, who said of her committee, "We won't be allowed to die until we have vomited enough."

As revealing in their own way were the outbursts of real aggression against people, not of the scientific élite, who presumed to comment favourably on the original idea of a moratorium as a socially responsible act, presumed to correct the rewriting of history that had already set in, to raise the question of germ warfare, to point out that lab workers had a different stake in research than that of the lab directors who employed them or to suggest that the clumsy, vacillating exclusion of the press had been a silly idea. All these points came up and in each case some distinguished scientist took his turn to put the boot in. This was after all a private meeting with an entry fee of £100; a little violence was likely to go unreported—and so it did.

What did emerge in the media reports was a split in scientists' thinking over the basis on which their judgements rested and how shifts in judgements on the safety question should be presented to the public. One faction felt no sense of unease in a dramatic change in the evaluation of the likely risks, from possibly large to certainly trivial, and believed that the public had to accept that. As Bob Pritchard, professor of genetics at Leicester University, put it, to loud applause, "There is only one thing that the public deserves and that is the truth as we see it." The other faction, including amongst its members Mark Richmond, at that time professor of bacteriology at Bristol University, found that stance politically naïve and based on a serious exaggeration of the reliability of the technical judgements. Political acts, like the moratorium, just couldn't be erased *that* quickly.

The cause of this collective frenzy was the idea and practice of regulation. Molecular biologists felt saddled with essentially unnecessary rules of conduct that bore no relation to the actual hazards of their work as they conceived them. Some of them clearly felt that their community had been ill-served by the initial gesture of a moratorium. Had that not happened, they argued, then the political pressures to construct specific regulations would never have arisen.

No wonder then that James Watson and other signatories of the Berg letter appeared to feel so guilty. When they were told that the action was praiseworthy they gnashed their teeth in desperation and roared with frustration. For the problem was how to get out of this predicament and abolish the regulations they had earlier called forth. How could scientists execute such a dramatic *volte-face,* whilst claiming first that the judgement that their work was potentially hazardous and then, four years later, that it was undeniably safe were both rational, authoritative and unbiased?

What happened was a gradual programme of relaxation, accelerated by occasional blasts from disgruntled researchers and from industry and assisted by constant comparison of the relative strictness of the regulations in different countries. In the UK the machinery of regulation, based on advice to the Health and Safety Executive from GMAG (the British Advisory Committee), has survived, with a major change in the way that the hazards of working with recombinant micro-organisms are conceived and devolution of powers to the laboratory level. In practice a great deal of laboratory gene-splicing now requires no special precautions, and many scientists have clearly stopped thinking about whether their research poses any special risks.

On GMAG's sixth birthday the infant was said to be in poor health, and there are some signs that this experiment in the social control of emergent research areas will soon be terminated. It was never an exercise in participatory democracy, despite the occasional leak. Just what the "public interest" representatives were supposed or permitted to do has never been made clear, although some of them have written at length about how they see their role.

GMAG and its relatives in other political cultures have been experiments. There are at least three different perceptions of them. For scientific administrators and government officials they were a convenient device for reining in contending parties and condensing disagreement into something useful in a situation of uncertainty. They provided a low-cost solution to the problems of political polarisation and could be used to assuage and incorporate dissent. They violated no important conventions of political behaviour. They were precedents but not disturbing ones. They allowed uncertain data to be translated into policy. This view was, I am sure, shared by the more patient and politically experienced scientists. Regulation in this mild form was a price worth paying for continuing research support, if it could be sold to scientists at the lab bench.

To many researchers, less schooled in opinion management and political machination, GMAG and bodies like it came to seem an affront to rationality and a ludicrous extravagance. Surely, some argued, a blank refusal to cooperate would serve to destroy its legitimacy and allow a rapid return to minimal regulation.

To a fairly small group of scientific radicals and trade unionists committed to social change, GMAG appeared in a third guise as an opportunity to gain a measure of public participation in the planning of research. If nonexperts could be allowed to take part in assessing the risks of science, surely they could do the same for its benefits. GMAG then was seen as a foothold, a place for those excluded from power to demonstrate the future need for their inclusion. This hope has, I think, been completely overwhelmed. Instead, GMAG has been carefully limited to the consideration of safe laboratory and industrial practice in those areas of biotechnology involving genetic manipulation. With effortless bureaucratic skill the radicals on the UK and US Advisory Committees have been penned in this corner, away from decisions about how money is spent on science. In any case the focus of the debate about biotechnology has shifted away from questions of risk to the promotion of innovation. Now the dominant issue is not, Are these experiments safe? but, How can we organise these experiments so that their commercial value is realised as soon as possible?

It is a remarkable trajectory in just under ten years. And for once, the ballistic metaphor is quite appropriate. At Wye College in 1979, a phrase that came up again and again was, "We have a re-entry problem." Apparently the biologists there felt themselves to have been fired into the high stratosphere of public debate, where they had been tumbling in orbit for several years. Now they wanted to return to the familiar state of scientific isolation. The problem was to judge the angle for their craft to slice through the thick veil of public scepticism. Cut it too fine and you are back in space again, debating the value and safety of your research with people you are quite sure are morons. Go too quickly and the social friction brings your mission to a disastrous end. But by 1979 there was already a great deal of money resting on getting these political judgements right.

When COGENE held its second international meeting in Rome in 1981, the agenda had changed radically. The major item under discussion was the problem of how to manage the impact of commercial pressures on university research. Funnily enough, many of the same people who had said that the risk question could be handled, were also of the view that commercialisation was really no problem either, as long as there were some rules of conduct and everyone agreed to stick to them. After all that approach had worked with biohazards, hadn't it? Of course, with safety regulations, things had started out on a very strict and cautious basis. Then, gradually, the rules had been lobbied, bent, forced and revised into triviality. Surely the same thing could happen again, couldn't it?

Biotechnology and the Economic Sunrise

Since the heyday of the debate over biohazards in the late 1970s, when Mayor Vellucci of Cambridge, Massachusetts, spoke of seven-feet-tall monsters emerging from the Boston sewers, several things have changed. Most notably, many molecular biologists, including some of the signatories of the Berg letter, have gone into business, either as industrial consultants, entrepreneurs, or as the hired gene-splicers of much-publicised genetic engineering companies.

These concerns give the biotechnology scene its special character, and they deserve an introduction in their own right. Although in the heart of this book I shall be writing about the plans and activities of enormous industrial corporations, the state of the technology at the moment means that an important role is being played by much smaller, nimbler organisations, bursting with ideas, talent and skill. These companies make research their business. They sell expertise in product development using the latest ideas and techniques in genetic engineering. They prosper by leading established companies through a new industrial maze, where it is said immense profits will be found.

Because high-powered research in cell and molecular biology is so fundamental to their operation, they have tended to spring out of university or government laboratories at the initiative of an enterprising individual or group of researchers. Typically they are founded on expertise with an experimental system or a technique. Some of the more successful businesses have moved off the campus and are growing at a phenomenal rate. A few have been floated as public companies, worth hundreds of millions of dollars. The names of the more successful or controversial—Cetus, Genentech, Biogen, Genex, Celltech, Agrigenetics, Transgene, CalGene and so on—will become familiar as we go on. This form of research activity, drawing together the excitement of discovery and the supposed glamour of a new, highly specialised industry, has its own mythology, which says that to be a genetic engineer is to be a pioneer in business *and* in science.

Just about every leading molecular biologist in the United States has some form of industrial consultancy, financial investment in a new biotechnology or direct salaried involvement. In Europe, much the same thing is occurring, although there are fewer companies being founded on campus by academic entrepreneurs. Nobel prizewinners from 1962, 1971, 1972, 1979 and 1980 have joined the game. I suppose that if you have already got a Nobel prize the chance of getting another must be infinitesimally small, although Frederick Sanger got his second in 1980 for DNA work. The prospect of merely training other researchers for the rest of one's career must be boring compared with that of making a pile of money for one's retirement.

For leading scientists, the pattern seems to be occasional consulting, leading to closer involvement in a sabbatical year, followed by full-time employment in a biotech company after resigning one's university post. For less exalted scientists, the move to industry comes after getting a doctorate or after several years postdoctoral shuttling around an international training network, picking up ideas, skills and contacts.

The attractions are not hard to discern, if you ignore the predictions that say that four out of five of the new small firms will go bankrupt. One scientist, Christian Anfinsen, Nobel prizewinner in 1972, resigned his job in the US to work in a Wall Street–backed operation in Israel, only to find that the plans had fallen through before he arrived. It sounds as if the company crashed while his plane was in the air. But that aside, firstly there is the money, which compares very favourably with academic salaries. If one is wanted badly enough to be offered a stock option as well, then there is a reasonable chance of considerable capital gains in a few years. Then there are the facilities, the palpable sense of excitement in a growing industry and the freedom from teaching and administration.

Since molecular biology is an intensely competitive subject it is doubtful that people work any harder than they were wont to do in academia, although there are stories of prodigious feats of scientific "yomping" to crack a particular technical problem. There are also stories of financial bonuses to go just that little bit faster. One man I met regularly started work at four a.m., often working through the night as well, although he had also done that as a university scientist.

Finally, it is clear that, at the moment, the genetic engineering industry is keen to nurture a research ethos, with relatively free communication of results, publication in academic journals and seminar programmes with visiting academic speakers on company premises. For a young scientist at the corporate lab bench looking back at academia, things must appear really quite similar, except for the cash. But at one recent meeting for potential investors in biotechnology, one grizzled old tomcat warned the youngsters that these halcyon days would not last. The time will come when gang warfare will break out between the different concerns and each scientific family will, like a Mafia one, "take to the mattresses" for a long war. Then, perhaps, the full burden of loyalty to a corporation may be felt and the research tasks brought down from their present grandiose level.

Looking in the other direction from university to Bug Valley, the view is significantly different. There is real anxiety amongst academic researchers about the antics and morals of former colleagues, now immersed in the business world. The worry is that ideas and results may be snapped up by corporate scouts and turned rapidly into dollars or some other currency, behind a screen of patent protection. It is not that communication in science was ever completely candid, at least in recent years, but by and large if another group wanted to repeat someone's experiment, and to build on it in their own way, they felt free to ask for the bacterial mutants or viruses or enzymes involved, and they expected to get them, perhaps after a slight interval as papers were readied announcing priority of discovery or the next round of experiments set up. Not everyone was a gent.

There is a story of one man, now a big cheese in genetic engineering, refusing to send a particular virus to his competitor's lab—a breach of normal academic protocol. The competitor claims to have cultured the virus from his letter of refusal anyway. Perhaps it was deposited there when he signed his name and put the sleeve of his lab coat on the paper. Generally the old system worked because of the threat of censure through professional institutions. Journal editors particularly were in a position to insist that the publication of papers was based on the understanding that experimental materials would be available on request. Various journals have been forced recently to remind their readers of their requirement, because commercial pressures are creating new habits of secrecy, lying and theft. Consequently industrial visitors are not always that welcome back on the campus, and one hears stories of notebooks locked away, results written in code and people out to lunch when former colleagues return.

Similar problems surround scientists holding university posts, whilst spending significant amounts of time building up their own commercial laboratories. At the University of California at San Francisco a determined campaign was mounted by biologists to get Herbert Boyer to remove his commercial activities from his department on campus because it was felt that his work for Genentech placed academic relationships under too much strain and distorted the balance of research in the department.

At the University of California at Davis, a plant molecular biologist, Ray Valentine, was asked to modify his relationship with CalGene and the Allied Chemical Corporation. CalGene is a biotechnology company set up by the son of a Silicon Valley millionaire. Allied Chemical awarded Dr. Valentine's group a $2.3 million research grant *and* had agreed to purchase a 20 per cent, £2 million share in CalGene.

It would be difficult, and possibly not desirable, to enforce a ban on commercial research on campuses. However, the delicate balance of trust and respect, or at least dynamically stable envy, between competing colleagues is all too easily broken by people being less than candid about the extent of their commercial activities and their interest in others' research. Inevitably people are secretive about how much money they are earning on the side, not least because they may be asked to share it with contributing students, secretaries, technicians and junior colleagues, not to mention their employing institution or research sponsor who, most likely, will have borne some of the overheads.

A recent editorial in the journal *Nature* suggesting that academics be required to declare their commercial links on a public register is a step in the right direction, but that system does not get to the heart of the matter, which is ripping off other people's work. It has been said that, for example, conventions of citation and acknowledgement in published work—the mannered but still necessary courtesies of a professional which indicate and sustain mutual respect—have changed recently. Establishing a claim to a patent may be complicated if too many people are represented as having made a particular experiment possible. Basically the fewer people you cite, the less you have to share the spoils. It is not that research supervisors have never published students' work as their own before, or heads of department never demanded undue slices of credit. It happens all the time. But now the sudden appearance of large amounts of money has increased the temptation to grab others' ideas and so increased the bitterness when that happens.

An example of the kind of misunderstanding that can arise in this way is the argument over the use of interferon-producing cell lines, created by two researchers at the University of California at San Francisco. Examples were distributed by them to colleagues and then passed on, in circumstances that are somewhat unclear, to commercial researchers at the drug firm Hoffman La Roche and its contractor, Genentech, which was working hard to produce interferon, a possible anticancer drug. The university felt that its property, the cells growing in culture, had been wrongly and covertly passed on to Genentech, which thereby gained a significant financial advantage, and this matter was taken to the courts. Whether it was sharp practice, carelessness or genuine misunderstanding I am not in a position to say. But it is worth noting that the aggrieved party was an institution with the resources to risk taking matters to the courts and not a lone individual.

It is possible, of course, to see all this as the price of progress. Friction of this kind, one might say, is likely to be transient, minimal, educative and not seriously damaging. Furthermore it simply points to the need to use the patent system as an established means of protecting intellectual property and for academics to be more energetic in seeking patent protection than they used to be. But on the other hand erecting a superstructure of patents could only be adequate if everyone concerned had the resources to defend their patch. Evidently that is not the case at the moment. At one recent business seminar on biotechnology, patent attorneys pointed out that patent litigation can be used as a form of economic warfare against vulnerable competitors. As a profession patent lawyers are likely to work on both sides of the street, advising some clients on how to use patents as protection, and others on how to evade or challenge them. Nice work, if you can get it. Patents, then, may be a mixed blessing. We shall come back to them later in the book.

The kind of entrepreneurship going on in biotechnology at the moment draws its appeal from the myths of the frontier, the genius-inventor and getting rich quick. The material basis of this phenomenon in the United States stems from the tax laws. Capital can be channelled into small businesses exploiting a few inventions or a spin-off from other research. Big companies like Monsanto or International Nickel set aside some of their earnings to back small ventures of this kind. The major banks, finance houses and stockbrokers also specialise in navigating corporate, institutional or private funds into research-based tax shelters, where investment in a series of new ventures at a higher risk than involvement with established firms brings an accept-able return from the aggregate of bonanzas and bankruptcies. As Mrs. Thatcher put it when the arithmetic was explained to her at Genex, a biotechnology company a short drive from the centre of Washington, "It's as exciting as betting on horses." As with horses, if you study form, it is possible to turn a disastrously expensive hobby into a reason-able income, although you can't entirely eliminate the risk. Capital brought together for speculative investment of this kind is called "venture" or "risk" capital. How it is used can vary somewhat, de-pending on whose money is involved and whether it is an unknown entrepreneur with limited resources, a world-renowned finance house or an industrial corporation "with deep pockets." The founder of Genentech, Robert Swanson, was in this business with the American bank Citicorp and, faced with the prospect of being sent by them to South Korea, opted to strike out on his own, recruiting Boyer as his associate. Biogen, another biotechnology company, was created by the man who ran this kind of operation for International Nickel. The Genetics Institute, set up by Harvard biologist Mark Ptashne, has funds from the personal trusts of the Rockefeller and Paley (CBS) families. Lord Rothschild, scion of the banking family and sometime research biologist, is now running a trust fund based in Jersey called Biotechnology Investments Ltd. The Prudential Assurance Company has, unusually for a large financial institution in Britain, set up a com-pany called Prutec to place money on technological bets of this kind. Stockbrokers McNally Montgomery have created a tax shelter for their clients, under the terms of the 1981 UK Finance Act, whose cash is invested in Cambridge Life Sciences, a company setting up to make money from the bacterial production of an enzyme, urokinase, that breaks up blood clots.

There is then a whole financial subculture that provides venture capital for "Route 128" entrepreneurs. This is the highway out of Cambridge, Massachusetts, where the small companies exploiting the spin-off from radar, micro-electronics and computing began to cluster in the 1960s. Like any other complex phenomenon it has its own dynamic and characteristic pattern of development. One American industrialist, obviously used to making use of these concerns, has described the process as follows.

> It begins when the founders get an idea for the company. They put up a very little bit of money. In Stage II they go to what we call a lead venture investor. That is usually a venture capital company. Depending on the idea and the needs, from half a million to a million dollars will be invested at Stage II. That takes the company through the first couple of years. Then, more capital is needed—Stage III. Now they go to additional venture investors or to large companies like Dow. The company has been in business for a while, perhaps we have a research contract with them, and we have a good warm feeling about them. That is the time when we would probably put in money. The money is more significant and the chance of failure is high. Finally the great moment occurs when the new company makes its first public stock offering.

One implication of this pattern is that a considerable number of companies will go bust, have to contract in size or get taken over by larger companies. Southern Biotech has indeed folded up, Bethesda Research Labs laid off 150 staff in 1982 and DNAX, Collaborative Genetics and New England Nuclear have been bought up (which is likely to have been lucrative for the founders).

Another implication is that publicity about research achievements will be carefully orchestrated to build confidence in a company, just before key moments in this process, such as public flotation. In an article in the *New England Journal of Medicine*, Spyros Andreopoulos, the press officer at Stanford University, claimed that in 1980 Biogen and Genentech had publicised results not yet authenticated by publication in a professional journal for exactly this reason. His argument was if this kind of thing was allowed to catch on, all kinds of dubious statements could be passed off as the truth. Claims like these are sifted out by the prepublication scrutiny that professional journals carry out. The line between practices that are routine in commercial hyping and the scrupulous procedures of academia is becoming hard to draw. In 1981, the public offering of shares in Genentech and Cetus Corporation, both Californian biotechnology companies, saw amazing scenes in the New York Stock Exchange. The price of stock in Genentech rose from $35 to $80 in the day's trading. This reputedly made Boyer a fortune on paper of some $50 million. Somehow nobody ever says how much money the other people involved made. It is Boyer that is the cult-hero. We may assume that at least the key executives and scientists in such a company will have become millionaires with a successful flotation, and they will have sufficient money now to buy the advice on how to look after their newfound wealth. Out of such experiences a potent mythology can be made.

However, after the heady events just described, the financial trends in biotechnology have been downwards for the most part. Significantly, leading journals now provide more and more financial data on the new industry for their scientist readers, since they are thought to be interested in the money markets. *Nature* indeed carries a regular table of share prices of biotechnology companies, compiled by Wall Street stockbrokers.

Biotechnology offers exciting prospects. It may seed new industries or rejuvenate mature, crisis-torn, established ones. In the forcing houses of the small genetic engineering companies splendid new industrial organisms may be brought forward. It may, like micro-electronics, satellites, minerals and ceramics, be part of the long-awaited economic sunrise, as another cycle of economic decline is left behind. That too is part of the mythology, the notion that genetic entrepreneurship could be a road to salvation.

This prospect is at least tinged with plausibility, and it has had a powerful effect on governments, looking for technological winners to back and concerned not to miss an area of technological promise needing financial support and other kinds of stimulation. Between 1974 and 1982 the governments of West Germany, Japan, the UK, France, Belgium, Canada, the US, the Netherlands and Eire have all commissioned reports on biotechnology, as have the European Commission and the Organisation of Economic Cooperation and Development, the economic think tank of the West. In addition, there have been government hearings and promotional policies designed to move things along. In the US, legislators have spent a lot of time considering various aspects of the innovation process and have pondered several bills intended to make life easier for biotechnology companies and to spur them to renewed effort by requiring government agencies to set aside money for them.

In a sense, the most striking and important of such measures was the decision of the US Supreme Court in 1980 that patents could be obtained on micro-organisms, and indeed on any species of living thing, provided that it could be shown to be a product of manufacture. This question was manoeuvred in front of the Supreme Court by companies with an interest in genetic engineering. If micro-organisms with new genes spliced into them and other engineered cell lines could be patented, as man-made, living things, then the field would be safer for serious corporate involvement. We consider this issue again in chapter 3.

But for all this legal and economic assistance the actual tasks of getting new products on to the market and building a secure financial basis for biotechnology have been proving more difficult than the mythology of research-led success allows. In the summer of 1982, the *Sunday Times Business News* carried a full page article with the head-line "The Gene Machine Runs Out of Steam." The implication was that financial splurging does not make a technological revolution.

Any investment is judged by the rate and size of its return as compared with other opportunities, taking account of inflation, tax laws and other accounting factors. Biotechnology is competing with, say, gold or property for speculative cash. But it turns out that the "payback period" is likely to be longer and the technical problems of going into production more complex than many people were told. The sun will rise slowly and the sky at dawn may be overcast.

The pause for thought, inspired by a return of investment caution, seems to me a valuable opportunity for wider public debate about what is happening. I am raising issues for public consideration at a moment in the process of innovation when the flood tide has slowed down. As a society we need desperately to consider what kind of future is being constructed through the operation of the financial, industrial and research systems. The biotechnology hype sounds a little overblown at the moment. Without wishing the tide of develop-ment to turn and ebb, we should ask what kind of goals form the basis of the industry into which this cash is flowing. The following chapters take up that question, followed by the industrial agenda I described above. In the final chapter I consider what alternative paths we could follow if we wished.

3

The View from the Cell

In this chapter I try to spell out the science of genetic engineering. I assume no scientific training, but people who have read it say that it is still tough going. **If you find the chapter daunting, read through it quickly or skip it and return after you've read the rest of the book.** In the long term, people like me who write such books and nonexpert readers will have to find a common language that satisfies us both if science is to be democratically controlled. At the same time, people with a scientific training will find the first part of the chapter elementary and may want to skip much of it. I have signposted the points at which you might want to rejoin the text.

Nowadays it's a commonplace of biology that organisms are made of cells, even though this degree of organisation is not something we can perceive directly with the naked eye. It is only 150 years ago that biologists, turning the analytic power of the microscope on living things, first concluded that their basic functional units were cells. The term also has the same meaning, of a basic unit, when applied to a honeycomb, a prison or an underground revolutionary party. Cells, then, are elemental, and compartmentalisation into cells is common to all organisms. In the limit, of course, as in bacteria, the cell is the whole organism. At the other extreme of complexity and development are human beings, composed of hundreds of millions of cells, which are organised into tissues, organs and systems of coordination and control.

Cells are organised capsules of synthetic activity programmed to carry out chemical reactions and to synthesise new materials. These reactions are speeded up and ordered by catalysts, which facilitate particular processes while remaining unchanged themselves. In biological systems these catalysts are called enzymes, and they are protein molecules, constructed from genetic specifications. Existence at the cellular level involves the continual processing of materials taken in from outside, by operating on them with molecular tools, made according to a work-plan located in and controlled from the genes. By rewriting the genetic text we can now redirect the activity of these microfactories. Biotechnology is, in large part, making cells do new things.

This is easy to state as an abstract idea, but it immediately prompts the question, How do you actually do that in practice? Cells are incredibly small and intricate. The molecular structures within them are several orders of magnitude smaller still. How then do you write the genetic text if the print is so fine? I am suggesting that the technical ability to move genetic instructions between organisms, linked to sophisticated chemical engineering skills, is crucial to the present acceleration of biotechnology. If it is crucial, we need to know how this is so. At the end of the chapter I discuss the technology—called process technology—within which cellular microfactories are located. I also have more to say about the emerging ability to design and synthesise living systems and/or their components on which I touched in the first chapter.

Thinking Cellular

The first idea that we have to take on board is that organisms are subdivided into basic functional units, cells. In general their size is such that they can only be distinguished with a reasonably powerful optical microscope. They come in a range of sizes, but each cell of the same type will attain the same size. Single-celled organisms, like bacteria, which are the most important organisms in present biotechnology, are typically one to two thousandths of a millimetre across. Objects that small can only be seen with an electron microscope. In multicellular organisms like a human being, their size, shape, structure and organisation vary a great deal. An average plant or animal cell is perhaps a hundredth of a millimetre across, which is significantly bigger than a bacterium, but still extremely small.

Cells of a given type have specific functions. The tasks necessary to maintain and reproduce an organism of a given species are divided up in characteristic ways. In a dandelion, for example, some cells transport nutrients from the soil to the leaves, while other cells break them down and turn them into the fibres and tissues of the dandelion. Although all cells contain hereditary information needed to create new dandelions, only certain cells, which form the seeds in the "clock," are constructed as vehicles to spread that information. Others mediate environmental stimuli, like the sun coming up, and organise a response to it. In an octopus the range of tasks to be performed is different; in a human being it is different again. In each case though, from one cell, seed or egg, develops a complex, differentiated aggregate of cells, the adult organism.

Self-Assembly from a "Genetic Blueprint"

How is it then, that dandelion seeds develop into dandelions and not into octopuses, blackbirds or stick insects? How does like beget like? How are the characteristics of a species transmitted between generations? To answer that question nowadays we often take an analogy from industrial mass production and talk of organisms being constructed from a blueprint, that is, a plan or representation of the finished object which exists in the seed or the egg. It is perhaps a better analogy to talk of a "programme" or set of instructions that guides the development of a mature organism.

Thus we can say that a dandelion is created from a specific set of instructions present in the seed from which each plant grows. Every seed is programmed to make a dandelion. Similarly for human beings: we arise as individuals by the repeated division of a fertilised egg. The union of egg and sperm brings together a full set of instructions to make a unique individual, characteristic of our species.

Growth to maturity is through cellular multiplication, according to an initial blueprint, moderated, of course, by an endless series of complex interactions with the environment. Germ cells, then, are programmed to divide repeatedly for a limited time and to organise themselves into a fully functioning whole.

But unlike something made on the industrial shop floor, organisms are self-assembling: that is to say, they form themselves by taking in material from the environment, breaking it down, processing it chemically into components and adding it to their fabric. And unlike a computer, which has to be made and then programmed, germ cells are preprogrammed as they are formed. The blueprint is internal to the entity being made, and it requires no external agency to read and act upon it.

The inherited instructions that say, for example, "Make this blue eye pigment," or "Make this substance to break down that sugar," we call genes. Thus inheriting the genes for blue eyes from one's parents means inheriting the instruction that the cells of one's iris should make the chemical that colours them blue. The simplest organisms, bacteria, possess some hundreds of (or possibly several thousand) genes. That is to say, some hundreds of characteristics are specified genetically to make a copy of a bacterium. Hundreds of instructions are passed on in the genetic programme to make the next generation. There are even simpler noncellular entities, viruses, that make do with even fewer genes. They survive by hijacking the machinery of those cells they infect to make more viruses.

In the most complex organisms, human beings, the number of genes is vastly greater, in the range of hundreds of thousands, possibly even millions. Surprisingly perhaps, every cell in a multicellular organism contains a full set of instructions for the whole organism. But only those relevant to the specific tasks of that kind of cell are operative; the rest are silent and switched off. Thus the cells of the pancreas don't make eye pigments, and no insulin is made by the cells of the eye. Any organism exists as a result of the cooperative functioning of these specialised units, each endowed with specific sets of instructions for their role in the body's scheme of things, and organised into higher-order systems, such as those concerned with circulation or vision or the perception of pain.

The Architecture of the Cell

Despite the different tasks they have to perform, most cells have common structural features and a common set of components. At certain stages in the life of a cell one can discern within it a number of rodlike bodies, called chromosomes. It is on these highly complex structures that the genes are located. We now know that all genes, except for a few in some viruses, are made of the same substance, deoxyribonucleic acid (DNA). DNA is a complex, immensely long molecule with an interesting and elegant structure, which consists essentially of two helices wrapped around each other and tied loosely together with weak molecular bonds. The structure of DNA is the famous double helix, proposed by Watson and Crick in 1953.

In the chromosomes of higher organisms, these helical chains of DNA are coiled again around beads of protein called histone, and these histone-bearing chains are then coiled again upon themselves into what is called a superhelix. At least, this is what scientists believe at the moment. This picture of chromosome structure might seem horribly complicated, which indeed it is, but it all serves to pack a vast amount of DNA—and a corresponding number of genetic instructions—into a tiny space in the cell nucleus. Also, it is now thought that this complexity is the basis of a control system that ensures that the right genes are switched on or off in specific cells. It also shows, I think, how complicated the task would be of getting new instructions into organisms, of doing genetic engineering, had not the inventive activity of evolution created natural ways of carrying genetic information around, getting it into recipient cells and getting those cells to use it.

Structure and Sequence

From cells, let us move to a finer level of structure—that of molecules. Let us also consider a basic biological process, like oxygen transport to particular tissues. In higher organisms this is carried out by specialised red blood cells, and within those cells millions of molecules of a particular substance, haemoglobin, are involved in the movement of oxygen around the circulatory system. Four molecules of oxygen bind to each molecule of haemoglobin, and they are released in turn as needed in the circuit around the body. Haemoglobin is a globular protein. That is to say, it belongs to the class of substances known as proteins, and it is folded up into a kind of ball rather than being stretched out into a fibre as the protein in muscle is. All molecules of haemoglobin have the same highly complex, irregular structure, and it is this structure that is the key to its function. The molecule is a composite of an iron-bearing "heme" complex and four chains of "globin," which are built up in the cells of the bone marrow from genetic instructions. The globin chains also have a unique linear structure. They are built up from a series of molecules called amino acids, which are derived from food or synthesised within the organism from other chemicals. Protein construction is essentially the formation of a unique sequence of amino acids into what is called a polypeptide chain—a molecular chain of amino acids joined by so-called peptide bonds—which then folds up into a unique configuration to carry out a particular task.

Indeed an important part of finding out how a protein works is to analyse its amino acid sequence. This is called "sequencing." Nowadays, it is fairly easy to do; it can even be automated. But the analytic skills involved had to be developed over time. For working out the amino acid sequence of insulin, the first polypeptide to be sequenced, Frederick Sanger won a Nobel prize in 1958. That kind of patient dissection of molecules into an ordered set of fragments is his *métier*, and in 1980 he won a second Nobel prize for developing methods enabling DNA sequences to be analysed.

A major research theme in molecular biology—and one of its fundamental intellectual principles—is thinking about biological processes in terms of molecular structure and function. The approach is via a question of the form, What kind of molecule carries out this biological operation? The first task is to analyse the process in detail to identify all the substances involved and the particular role of each. Then its three-dimensional structure must be worked out. The task took twenty-five years for haemoglobin, which is a very complex molecule. It took another fifteen years to produce a satisfying explanation in structural terms of how the haemoglobin molecule functions. Nowadays, three-dimensional structural analysis is somewhat easier to work out. The amino acid structure—the one-dimensional structure—of a protein is very much easier. What Sanger did over eight years to win his first Nobel prize can now be the work of a day or two. It is now accepted that one-dimensional structure determines three-dimensional structure, that is to say, a molecule with a given (one-dimensional) amino acid sequence will fold up into one characteristic configuration, and only that one. But unfortunately knowing the sequence still does not allow one to deduce that unique structure. Folding is a very complicated business.

The Idea of a Genetic Code

In the 1930s, biologists used to think that proteins might well be highly regular molecules, complex in structure, but still neatly fitted together. By the 1950s it was already clear that this was far from true. Proteins are tangled. Yet their disorder is only apparent: they are like an incredibly elaborate knot, that will always form the same shape, if you follow the same set of operations, unlike for example a plate of spaghetti, which will never pile exactly the same way twice. So how do molecules like that get built? Consider, for example, that in making one red blood cell, 260 million haemoglobin molecules have to be put together, all of them the same. The answer is from genetic instructions, that are followed faithfully over and over again, to order amino acids into the sequence characteristic of haemoglobin. Yet what form can these "instructions" take? The answer that emerged with clarity in the early 1950s, though it had been suggested rather speculatively twenty years earlier, involved the idea of a "genetic code." Sanger's work showed that to construct insulin molecules *sequence* information had to be provided. Molecular biologists reasoned that presumably the same would be true of all other polypeptides. Somehow the genes must specify a sequential or linear order. Yet genes are made of DNA, which is quite different chemically and structurally from protein. So genes could scarcely order proteins by acting as a pattern or template, but if the connection was formal, if some structural feature of DNA represented or *encoded* a particular amino acid, then a sequence of such configurations in the DNA could be read as a unique protein sequence. If cells could read information in DNA and decode it as protein, then the problem of how proteins are specified was solved. One of the exciting things about the double-helical model of DNA was that it showed immediately how to push this analysis further.

Positioned at the centre of the DNA molecule are the four members of a select set of chemical units: adenine, represented in the text henceforth as (A), thymine (T), cytosine (C) and guanine (G). These are the four letters of the DNA alphabet—the code of life. Their structure dictates that they must always form complementary pairs. A only bonds to T, and G only to C.

Each pair of bases is joined together by relatively weak chemical bonds, called hydrogen bonds, which split apart fairly easily. The point is that to make a new generation you need a new set of instructions. When the DNA molecule separates into two strands, each acts as a template for a complementary strand, thus forming two new double helices. If that process of forming a new generation is to be easy, then the DNA must separate easily. Other molecules tend to be stuck together more securely.

Thus within any DNA molecule one finds at the centre an enormously long string of paired bases. We now realise that these sequences of chemical components form messages written in a simple code. When the bases are taken in groups of three, called "codons," they stand for particular amino acids, or for specific instructions to be acted upon in the synthesis of a polypeptide chain. A sequence of codons specifies the order of a polypeptide chain. Each of the sixty-four ways of forming a set of three elements out of A, T, G and C, such as ATT, GAC, TAT, etc., has a meaning. Most of them stand for a particular amino acid. A few of the coding triplets are signals which say, in effect, "Message ends here; stop adding amino acids to the chain already formed" or alternatively, "Message begins here."

The genetic code is, then, the set of relationships that link two alphabets, the DNA alphabet and a protein alphabet. If this isn't clear, think of Morse code as an analogy. In Morse, the encoding alphabet has just two symbols, "dot" and "dash." To encode English you form groups of dots and dashes to represent each of the twenty-six letters of the written alphabet. The genetic code is neater, in a way—the coding groups are all triplets of the same size—and more complicated, because different groups of symbols (codons) can stand for the same thing. But its function is the same, to relate messages written in two different alphabets.

Genes Order the Synthesis of Proteins

Making a protein then is a matter of reading out the encoded information in chromosomal DNA, and, with the assistance of a variety of enzymes and other helper molecules, getting amino acids assembled into a specific sequence. A number of genes are involved, apart from the one that codes for the protein concerned. Various genes code for enzymes that stabilise and facilitate the synthesis of the polypeptide chain and assist the reading out of the genetic information. Others code for the adaptor molecules called "transfer RNAs" that pick up particular amino acids and move them into position. Yet other genes form part of a control system that ensures that the process stops when enough protein molecules have been formed or that synthesis starts up when the need for a particular substance arises.

Protein synthesis occurs essentially in two stages, the first in the nucleus, the second in the surrounding area, or "cytoplasm." First, an enzyme called "RNA polymerase" binds to a specific location on the DNA molecule at or near the beginning of the gene to be expressed. As the enzyme moves along the DNA strand, a complementary copy of one of the strands is made out of *ribonucleic* acid (RNA), the DNA acting as a template for the RNA. RNA is very like DNA chemically, except the deoxyribose sugars are replaced by ribose sugars and the base thymine in DNA is replaced everywhere in the RNA by the base, uracil. RNA in this role, which is called "messenger RNA," is single-stranded. Messenger RNAs are intermediates between the genes and the cytoplasm. They are *transcriptions* of a set of instructions written in DNA language into an intermediate RNA language, which is then *translated* into protein.

In 1977, it was discovered, much to biologists' surprise, that genes in higher organisms like chickens, frogs, fruit flies and human beings, unlike those in bacteria, are punctuated by sections of DNA that do not code for any part of the protein specified by that particular gene. Just what these noncoding intervening sequences, or "introns," do is a mystery. Most genes are punctuated with stretches of nonsense, and this has to be edited out between transcription and translation. Before the RNA molecule passes out of the cell nucleus, where it was made, the intervening sequences have to be cut out. The remaining sections of the molecule that do have a coding function are then rejoined in exactly the right order.

The edited version of the messenger RNA diffuses through the membrane around the nucleus into the much larger, outer region of the cell called the cytoplasm, where it binds to a body called a "ribosome." Most cells have many thousands of ribosomes. It is on them that free amino acids, linked to special adaptor molecules of another kind of RNA, called "transfer RNAs," are manoeuvred into the sequence dictated by the messenger and thereby added to a steadily lengthening polypeptide chain. It is as if they were queuing up to be in the right order. As each unit is added, the messenger RNA moves through the ribosome, rather like the tape on the reading head of a tape recorder, until the sequence is complete. The ribosome is, as it were, the recording studio. The newly formed polypeptide chain then folds up into its normal shape, possibly after some further processing with bits snipped off and stuck on here and there. The ribosome is by now free so that another message can be transcribed. It is thought the whole process takes several seconds to complete and that cells in top gear can turn out several thousand protein molecules of a given kind every second. The whole thing represents an extraordinary feat of coordination and microminiaturisation, if you think that a hundred billion bacteria could fit into a teacup with ease and each cell lives by producing hundreds of different proteins, many of them being made simultaneously, in controlled quantities at the right place in the cell. So the elements are these:

DNA provides the plans—specified in the triplet code of base sequences.

Messenger RNA and transfer RNA act as intermediaries.

Amino acids are the raw materials which are put in order—intermediate chains are polypeptides, and long chains are proteins.

The plans are handed over to the intermediaries in the cell nucleus.

The amino acids are put in order on the ribosomes.

This then is how cells grow, by building the molecules needed for their own structure, and how they perform their specific metabolic tasks. They create the substances and catalysts needed to process ingested material, turning it either into stored energy, or molecules to be stockpiled for use elsewhere in the body, or waste. After a while cells begin to organise their own duplication. The material of the cell is subdivided and each chromosome doubled until two identical cell copies gradually become distinct and separate.

The world of cellular synthesis is hidden from our senses by the barriers of scale. At the ultramicroscopic level, a level that we can only just reach with the electron miscroscope (and to do that we have to freeze everything in time), this dizzying pattern of activity is going on ceaselessly, and with incredible coordination. To talk of the cell as a factory, as I did earlier on, is a considerable compliment to human industrial organisation.

Given the intricacy and delicacy of the systems described here, perhaps the remarkable thing is that we can actually redirect cellular production. Genetic engineering is in essence the establishment of a degree of control over the processes described here by inserting new instructions into specific cells, either to correct a functional defect, or to make the cell perform a task that hitherto it could not have done. Since genetic instructions are encoded in the base sequence of DNA, this amounts to inserting specific DNA molecules into host cells, in ways that complement, or dominate, but do not obstruct, their functioning as cells. Our ability to move genes around and to reprogramme cells is based on the novel use of tools and techniques already employed in nature.

Genes on the Move

In industrial biotechnology we are usually talking about inserting new genes into bacteria or other single-celled organisms like yeast. In medical genetic engineering the task is made more complex by the need to locate and alter the genes of a specific group of cells, perhaps in one organ or tissue, and to affect only those cells.

We start with the simpler problem, playing around with genetic programmes of bacteria. Bacteria have been the preferred organism of one group of biologists for forty years. They are cheap and easy to grow, they are extremely small and a new generation appears every half an hour or so, rather than every week or month, and as organisms they are relatively simple. Since there are thousands of bacterial species, and within each species there are thousands of variations on the basic pattern, scientists interested in bacterial genetics have tended to concentrate on one strain, with the result that a great deal is known about its genes. The micro-organism in question is called *Escherichia coli* (*E. coli* for short) and the strain in common laboratory use is called K12. There are hundreds of *E. coli* strains. Some live in the human colon (large intestine) where most of them are harmless, although some can be very unpleasant; others live elsewhere, such as in wounds or in the tissue surrounding the brain, where they can cause big problems. *E. coli* K12 is a very inoffensive, even rather puny, creature, and it will grow happily in the laboratory, away from its tougher, very competitive relatives, on a diet of a few sugars and minerals, spreading out on a layer of jelly in a round glass dish. It can also be grown in a liquid environment, in a fermentation vessel. Either way, with enough food, every twenty minutes each bacterium will divide itself into two, which soon leads to an enormous number of micro-organisms until, as must always happen, the food in the surrounding medium has all been used up.

Bacteria, just like any other organism, exist by exploiting the resources of their environment, whether it is in the human intestine, in cow's milk, in the root hairs of a soya bean or a sulphurous hot spring. These resources, being finite, will allow only limited populations to survive. Like any other organism, bacteria are the products of evolution, and the capacities they possess are the legacy of interactions with other life forms, including the organisms they attack or feed off, those with which they establish some kind of collaboration, as do *E. coli* with humans, those with which they compete and those that attack them, such as viruses.

For our purposes here only one evolutionary property of *E. coli* need be explored—the ability to transmit items of genetic information between individuals and strains in a population. Bear in mind that, in general, bacteria reproduce by splitting in two, i.e., nonsexually, so that the new generation is an exact copy of the old, barring the odd mistake. If they all did that all of the time then the appearance of any new properties would be exceedingly slow. It is new properties, arising out of the trial of many possible variations, that are essential to staying in the evolutionary competition. We know now that bacteria can engage in sex, once in a while, and even that can't do much to lighten the tedium of being a bacterium. But at least in sexual union, or "conjugation" as it is called, sets of genes from two different partners are combined together, so that novel individuals appear.

Bacteria are also attacked and destroyed by much simpler noncellular entities, "viruses," the things that cause colds, measles, etc., which are usually just a long string of DNA wrapped up in a neat coat of protein. In response to this threat bacteria have evolved several ways of turning the attack and using the viruses as the carriers of bacterial genes between bacterial cells. This is what happens in "transduction," where a bacterial cell is entered by a special kind of virus called a "transducing bacteriophage" or "phage." The phage DNA, which is injected into the bacterial cell, carries a number of bacterial genes that are transduced, so to speak, or led across, from a donor bacterium from which the phage has been released.

Then again, some bacteria carry some of their genes on rings of DNA, called "plasmids," that are physically separate from the main circular chromosome. Genes packaged in this way can be more easily transferred to other bacteria. Thus a trait like resistance to a particular antibiotic can be rapidly passed on between organisms, either of the same strain or to other strains. The point is that out of millions of years of evolutionary struggle have come mechanisms such as these for passing on genes to other bacteria. Having such adaptive tricks on hand, and many more besides, has enabled them to survive in changing environments.

For some years phage transduction and bacterial conjugation have been used by biologists to manipulate bacteria genetically for research purposes. A great deal of research in molecular biology involves tinkering with the properties of bacteria, searching for mutants that lack a particular capability, which is shown thereby to be under genetic control, or trying to delete or block one step in a biochemical pathway to see how the complete system works. For example, bacterial conjugation allows one to cross bacteria, to set up an exchange of genes between "male" gene-donating organisms and "female" gene-accepting ones. The trick is to ensure that the hybrids can be discriminated from the parents, and this is often done by growing the bacteria in a medium in which only bacteria with the desired combinations of genes will survive. But these methods have their limitations. After all, you have to have the genes in bacteria already, before you can move them around. The techniques of gene-splicing allow much more radical gene transfers to be set up. Now why would you want to do that?

Readers well versed in the fundamentals of modern biology might like to start reading again at this point. Those who know a lot of recent molecular biology will probably want to skim one more section.

Putting It All Together: Gene-Splicing

At this point, it might be helpful to recall what I said in the previous chapter: in the late 1960s some molecular biologists felt that the golden age of their subject had passed. One of them, Gunther Stent, wrote a provocative semiobituary on the field, entitled, "That Was the Molecular Biology That Was." Indeed, the ground that we have covered very sketchily in the last six sections was the terrain covered by molecular biologists from 1945 to the late 1960s. In its own terms it had been extremely successful. A major revision of the fundamental conceptual framework of biology had been pushed through. One could now think of organisms and their physiological activities in terms of genetic information, code, structure, sequence, transcription, translation, feedback and replication. Francis Crick called the complete working out of the genetic code in 1966 "the end of the beginning."

Some molecular biologists felt that there were still a large number of interesting problems to follow up in the understanding of bacteria. Jon Beckwith, for example, was not forced out of front-line academic research by the conflicting emotions and commitments of being a socialist and a scientist, as were a number of his generation. Instead, he has continued to work on the way in which the secretion of proteins through bacterial cell walls is controlled genetically, and to discuss publicly the social and political uses of genetics. Ironically, his work now has an importance for biotechnology, although Beckwith has, on the whole, chosen to evade commercial involvement. Other molecular biologists moved into the study of higher organisms, and began to look at processes such as embryonic development in general, the formation of a nervous system, the immune response and the cellular transition to cancer. All of these turned out to be formidably difficult to analyse. The big question seemed to be, How would one ever manage to analyse, at the molecular level, the dense pattern of activity in the cells of higher organisms?

By the mid 1970s, however, it was clear that a major increase in analytic power was occurring. Several lines of research had converged to produce a new and immensely powerful methodology. Essentially, it allowed one to dissect genetic systems with much greater precision and sophistication and examine their operation piece by piece. To use the familiar programming metaphor, one could isolate particular instructions used in higher organisms and transfer them to a simpler programme being run on a less complex machine, i.e., a bacterium. By moving genes into bacteria, one could see more easily what they do in the organism from whence they came. But in the process of developing this research technique it became obvious very soon that reprogramming bacteria had other, industrial uses, and with that realisation the present era of biotechnology began.

This breakthrough derived from an ability to cut DNA molecules with great discrimination just where one wanted and to splice DNA fragments together. For that reason the techniques are sometimes referred to as "gene-splicing"—and the analogy with splicing together sections of magnetic tape or film is a close and helpful one. The result of such splicing operations is to recombine genes or sections of DNA. Such hybrid molecules are therefore called "recombinant DNA."

The First Cutting and Splicing of DNA

It is one thing to get a molecule into a cell, but if that molecule carries an instruction, the molecule has to be copied with each generation. One needs to get it reproduced and "expressed," i.e., doing its job, along with the genes of the recipient (or host) cell. Some viruses are very efficient at getting themselves into bacteria. They carry a special enzyme kit to help them break and enter, but once inside they tend to do a great deal of damage. So viruses have disadvantages as carriers of genes, if one wants to incorporate those genes into an intact, stably reproducing cell.

In the early 1970s, however, it was discovered that the cell wall of *E. coli*, when treated with calcium chloride, became permeable to plasmids—rings of genetic material on the loose within the cell. Under these artificial and easily reproduced conditions, plasmids would therefore enter the bacterium. Plasmids offered a more promising way for getting genes into bacterial cells as part of a system that would replicate inside the host cell, if there were some way of splicing them into the plasmids.

The Scissors

Around the same time, molecular biologists discovered a set of enzymes in bacteria, called "restriction enzymes," that acted as a kind of molecular scissors by chopping up DNA molecules into large, non-functioning fragments. They were the agents responsible for a peculiar phenomenon discovered in the 1950s whereby bacteria seemed to be able to pick up the hereditary ability to defend themselves against attack by specific viruses. The range of organisms that a particular virus could infect was "restricted." At first it was thought that restriction enzymes were another form of defence that had been created to meet the ever-present viral threat. It turns out that as a defence mechanism for bacteria, restriction is not very effective. It is now thought that it serves to police the boundaries between bacterial species, keeping them distinct by chopping up the DNA from one species that might have found its way into the cells of another. Overall a mechanism of this kind would help to speed up evolutionary changes, by acting against the blending and dissipation of species differences.

It soon became clear that restriction enzymes cut DNA only at certain sites defined by a particular base sequence. Thus the enzyme Eco R1, from the KY13 strain of the bacterium *E. coli,* was found to operate only at places where the bases GAATTC were to be found in that order on one DNA strand. On the other strand the complementary base sequence would be CTTAAG. If you write these one above the other, as a way of illustrating their complementary pairing more easily than with a picture of a double helix, it should be clear that

G A A T T C
C T T A A G

the sequence is a palindrome: notice the geometrical symmetry—it reads the same forwards as backwards, with an axis of symmetry in the middle. It was found that the enzyme Eco R1 broke the outer chains of the double helix between A and G, that G and C remained paired, and four bases, A, A, T and T, on each severed strand were left unpaired when the DNA molecule was cleaved.

G A A T T C
C T T A A G

Now the distribution of bases in chromosomal DNA seems to have no real order, so the chances of finding this particular sequence of six bases would be very small. Here then was a usefully fastidious enzyme that would only break DNA at rather uncommon locations. If, with luck, this sequence were to appear near a gene of interest, then one might be able to cut out the gene intact. All the enzymes previously known to cut DNA did so with such undiscriminating relish that one was left with thousands of useless pieces of DNA and no functioning genes. The number of known restriction enzymes has increased dramatically from the early 1970s to over 300. Each of them recognises a particular uncommon base sequence and breaks the DNA at the same or related point. If one knows the base sequence of a particular section of DNA one can break it more or less wherever one wants by picking the right enzymes. You can for example break plasmids open at just one spot, so that they can be rejoined by splicing in one piece of DNA. But how do you get DNA molecules to join or rejoin? You can add "sticky ends" of known composition to DNA molecules that lack them. By ensuring that the two sequences are complementary, their joining is assured.

So far then we have a tool kit for cleaving DNA at specific locations —enzymes extracted from different bacterial species—and for repairing DNA or adding short sections to it. Around 1972, biologists began to chop up plasmid DNA with these tools, to find out what fragments a particular enzyme will generate from a given plasmid. Generally speaking, a circular plasmid of, say, 5000 base pairs in length will have just one or two restriction sites for a given enzyme. Others began to explore the similar action of restriction enzymes on viral DNA. They immediately began to think not only of cutting but also of splicing. One necessary precondition for doing such experiments was to have some way of knowing whether or not particular genes—those spliced into the virus or plasmid supposed to be transporting them—had been taken into the target cells. Plasmids are "extrachromosomal DNA." They are small sets of genes separate from the main bacterial chromosome. One of the more common characteristics that these genes encode is antibiotic resistance. Since the discovery of antibiotics in the 1940s, we have discovered that bacteria can become resistant to them, and the more drugs we use the more likely that is to happen. Moreover, antibiotic resistance can be passed on from strain to strain. It is plasmids that spread the good (or from our point of view, bad) news.

Sticky Ends

When some restriction enzymes cleave DNA they do so in a way that leaves one DNA strand longer than the other. Unpaired bases, looking for a partner, remain on the longer one. These are called "sticky ends" because of their propensity to join up with other DNA fragments if they end with the correct complementary base sequence. This property can be exploited by using other enzymes that will add one by one the requisite bases to the end of one strand of a DNA molecule. Around the same time that restriction enzymes and their mode of action were uncovered, scientists also discovered other enzymes that repair broken segments of DNA by rejoining the bonds on the outside of the molecule. This class of molecules called "ligases" could therefore speed up the rejoining of DNA fragments, created by restriction, that were to be spliced together.

Labels

In gene-splicing experiments the phenomenon of drug resistance which can be inherited has turned out to be useful, since one can identify the cells into which plasmids, with new genes spliced into them, have passed by arranging things so that the plasmid confers resistance to a particular antibiotic on a strain of previously sensitive bacteria. Drug resistance could act as a label for a successful splice.

Here, then, were all the elements for a new mode of gene transfer: plasmids, capable of entering a cell, and being replicated in it; enzymes that could break up DNA and others that could repair it, allowing recombinant molecules to be created; and a method of detecting whether a particular gene transfer had taken place.

The First Splice

One experiment that demonstrated the practicability of this technique was carried out in 1973, through the collaboration of Herbert Boyer at the University of California, San Francisco, and Stanley Cohen at Stanford. It turned out to be the basis for genetic engineering.

They started by treating bacteria so that the chromosomal DNA and the rings of plasmid DNA were released from the cells into the surrounding medium. The plasmid DNA is smaller in size than the bacterial chromosomes and can be made to differ from them in density and can therefore be separated from them by spinning the two sets of molecules in an ultracentrifuge. The plasmid they used was a small one, known to confer resistance to the antibiotic tetracycline and called pSC101 (plasmid Stanley Cohen No. 101). At the time they thought that they had created this plasmid by mechanically shearing a larger bacterial plasmid. That turned out to be wrong, and in 1977 Cohen discussed in a paper where the plasmid might have come from. The significance of this will emerge later. In due course they sent other workers copies of the plasmid so that they could do similar experiments.

When the plasmids had been isolated, they were broken open by a particular restriction enzyme, specially selected so that it cut this molecule in only one place, producing linear strands of DNA terminated by sticky ends. They then allowed the plasmids to rejoin and repaired the DNA with ligase enzymes and persuaded *E. coli* sensitive to tetracycline to take up the plasmids. The bacteria acquired resistance to the drug and the plasmid was copied in subsequent cell divisions. The next step was to see if a foreign DNA could be spliced into pSC101 without impairing the functioning of the plasmid as a genetic element, or the expression of its genes. They mixed pSC101 DNA with another *E. coli* plasmid, that confers resistance to another drug, kanamycin. Some of the cells into which the resulting plasmids were introduced became resistant to kanamycin and tetracycline, suggesting strongly that the two plasmid fragments had combined, so that the bacteria were accepting two sets of genes spliced together.

Then they repeated the procedure with a plasmid from *Staphylococcus aureus,* another bacterial species which does not exchange genes with *E. coli.,* and found that a trait encoded on the *Staphylococcus* plasmid could be passed on to *E. coli.* Finally they set about splicing into pSC101 a gene from a totally different organism, a *Xenopus* toad, and found that the animal gene was indeed copied in generation after generation of bacteria containing the recombinant plasmid.

These then were gene-splicing experiments of arresting significance. They created "recombinant DNA molecules" or "chimeric DNAs" (named after the mythological Chimera, a creature with the head of a lion, the body of a goat and the tail of a serpent) by splicing together DNA molecules from different species. In some cases at least, the genes from one organism were expressed in the cells of another. The implication was that bacteria could take up instructions from other species, that the inserted genes could be replicated and these genes could be made to function in the host bacterial cell. If similar vectors could be found for other kinds of cells, then possibly plant, mammalian or human cells could receive new genes. But for a start, bacteria could be reprogrammed with extraneous genes. That in itself was a staggering possibility.

Not dissimilar experiments were planned during the same period by a group at Stanford led by Paul Berg, who wanted to splice together viral DNAs to explore the action of the tumor virus SV40 (simian virus 40), which infects and transforms monkey cells. This idea, which was also exciting, produced some concern since, after all, Berg was talking here about joining up part of a cancer virus with a bacterial virus, able to get into the bacteria that live in the human intestine. In a new cellular environment, what would the new hybrid do? Equally if those infected cells passed on their viral hijackers to more robust microbes outside the laboratory, what would happen then? Berg was sufficiently impressed by the uncertainties to defer his planned experiments and, as I have described, to organise a moratorium.

Prospects

These techniques promised an enormous increase in researchers' abilities to search for and analyse particular genes, which is an important part of the molecular biological enterprise. But there were some mouth-watering commercial possibilities. Thus, one could think about getting human genes expressed in bacteria to make a human protein. That was an amazing prospect, for there were all kinds of medically useful proteins, that are expensive to isolate for research purposes, let alone make in quantities large enough for therapeutic use, that might be made far more easily by reprogramming molecules with human genes to make them.

Patents

This did not escape Cohen and Boyer, who applied for a patent on this basic technique in 1974, within the year allowed under US patent law after the first public disclosure of their work. This was an astute, if unusual and controversial, move to make since, in 1974, such commercial-mindedness was not common among scientists. It was also an act of breathtaking arrogance to appropriate the property rights to an entire field of research and an emergent technology, created over decades by the communal endeavours of a profession funded by the taxpayer, and in the latter stages by teams of scientists and laboratory staff led by Boyer and Cohen respectively. It implied an appreciation by these men, or their advisers, of the scale of the likely industrial interest in this technique and of the feasibility of making a patent claim for this work. The patent may well turn out to be the most lucrative in history.

It has been suggested that the decision by Stanford officials to seek a patent marked a particularly clear realisation that, henceforth, universities would have to rely more heavily on commercially generated revenues from hitherto underexploited legal arrangements of this type. The law in the US was changed in 1980 so as to allow universities receiving federal research grants, as Cohen had been in this case, to seek patent rights on the results of such research, providing the revenues are spent on teaching or research. The universities' gain here is a loss to the state. The supposed justification is that this option seems more likely to speed up the rate of innovation.

The plan by Stanford University to patent Cohen's and Boyer's technique remained a secret until 1976, when a speaker at a symposium on genetic manipulation at MIT mentioned the persistent rumours then circulating that someone was trying to take out a patent on the basic techniques. Possibly he knew what was going on and wanted to bring things out in the open. Stanley Cohen, who was at the meeting, mentioned that Stanford and the University of California were exploring the patentability of his and Boyer's work. He also stressed that he did not stand to gain financially from this action, were it to be successful.

The pursuit of the patent claim on the *process* of genetic manipulation pioneered by Boyer and Cohen was not straightforward, and the application was reformulated twice. Not surprisingly, it also turned out to be a source of irritation for former collaborators, who were not being offered a share of the royalties. Taking a central role in the application was a paper written by Cohen and Boyer and two other people, Dr. Annie Chang of Stanford and Dr. Robert Helling, who in 1980 was associate professor of botany at the University of Michigan. The contention was that Boyer and Cohen had provided the essential creative input, and the others had been mere instruments of their inventive will. Helling was not pleased by this and refused to sign a form agreeing that his role had been marginal. At one point there was a real possibility of a lawsuit being brought by Stanford in order to depose him from any claims as a participant in the invention. Another scientist, John Morrow, involved in the work with *Xenopus* DNA, who had moved on to Johns Hopkins University in Baltimore, was also irritated by his exclusion.

In the event, these problems were dealt with somehow, and the Patent Office issued Patent No. 4,237,224 on 2 December 1980. The Office of Technology Licensing at Stanford then began to draw up terms for licensing arrangements for companies wishing to use these techniques, academic researchers being exempt from such obligations. They decided on an initial nonexclusive licence fee of $10,000 and an annual fee of the same amount for use of the techniques in research and development. Additional royalties on sales of products were set at one per cent for sales up to $5 million, falling to 0.5 per cent for sales over $10 million. This may sound like a substantial levy, but it is said to be relatively modest as these things go. The trick in constructing a scale of charges is to set them at a level that will bring in a nontrivial income, but which is not so much that a challenge to the patent would be worthwhile. By mid 1982, seventy-three firms had paid the initial fee, so that Stanford and the University of California between them must have collected some hundreds of thousands of dollars. By then Stanford had filed further patent applications, including a major one concerning all the products of processes involving this form of genetic manipulation.

At the beginning of August 1982, the US Patent and Trademark Office suddenly announced that the second patent was to be rejected. It is said that the reason for this was the uncertainty over where the plasmid pSC101 had come from. It seems that if one were to follow the procedures described in the first papers by Cohen and Boyer, and included in the patent application, one would not have ended up with the intended plasmid. Cohen had indeed considered other possible origins of his plasmid in an article he published in 1977. There is also a suggestion that prior publication of the results had occurred at a scientific conference, of which a popular but highly detailed account was published in *New Scientist* in October 1973.

The earlier patent had also been attracting critical attention from other quarters. Albert Halluin, a leading patent attorney employed by Exxon Research and Engineering, published a paper in August 1982 in which he identified four technical points of weakness in the way the application for the Cohen-Boyer patent was made. It is difficult to view this paper as merely an academic exercise, since Exxon does have research interests in this area, which implies longer-term commercial interests as well. There was for a time the possibility that the first patent would be withdrawn by the Patent Office, and a direct challenge to it still is possible. From one point of view that might be a good thing; on the other hand the revenues have been going to an academic institution. Had Genentech filed the claim then criticism would probably have been much louder. As we shall see in chapter 4, that kind of corporate grabbing of common intellectual property has indeed occurred. The fact that Stanford is an academic institution is said by some to validate its actions, which, on this view, are merely a sign of the economic times and an intelligent use of the legal system.

Effects of Patenting on Research

If this practice becomes more common, as of course it will, then there is a real danger that relations between colleagues, status in the university, criteria for advancement and recruitment and the allocation of resources for research will all be affected. One possibility is that purely commercial research will drive out the less immediately applicable, academically more promising work and that conceptual development in biology will slow down. Mere secretive tinkering with industrial bacteria will supplant fundamental investigation; for example, researchers will go for short-term profit instead of benefiting humanity.

The assumption here is that industrial research and development in general do not and cannot lead to fundamental scientific advances. Sometimes clearly they can, depending on the phenomena involved and the way the research is managed. However, the more important issue, in my view, is the question of which goals get built into fundamental and applied research. What we have to decide is not what institutional arrangements will keep the research front moving forward, but what institutional arrangements will allow research for social needs defined in other ways than by the operation of the market. How do we open up space for alternative programmes of research that set aside and confront the priorities of profit-maximising industrial corporations rather than merely protecting the traditional space for research élites to pursue their own preoccupations?

The question of patents has other dimensions, such as the issue of whether life forms like bacteria may be patented at all. The first Cohen-Boyer patent covers a process or set of techniques. But commercial interest in biotechnology has also forced consideration of whether specific organisms could be regarded as private property, if they were created by making gene combinations that never occur in nature. If this were possible, then it would offer a novel form of protection to manufacturers in the field. On the other hand, the patenting of life forms and the private ownership of species, albeit simple ones, struck some people as unacceptable. It was a challenge to the common view of the fabric of nature as something to be shared as a common heritage. If one could patent bacteria, why stop there? How about electrons? Or species of mice? Or cattle? Or people? Or all of the clones of a great horse, a great athlete or a great scientist?

In 1972, an Indian scientist, Ananda Chakrabarty, working at the Schenectady laboratories of the General Electric Company, filed a patent application on a specially created *Pseudomonas* bacterium that he had produced, without gene-splicing, which had an enhanced ability to break down four of the principal components of oil. A possible use might be to contain and remove oil slicks at sea, although the application had all the marks of a test case with no practical potential. Against the background of recombinant DNA research, it certainly came to be seen as a test case by biotechnology companies and by the US Patent and Trademark Office, which referred the matter to the Supreme Court in order to obtain a ruling on whether existing patent legislation, as enacted by the US Congress, seemed intended to permit the patenting of life forms. A five to four majority in the Supreme Court held, in June 1980, that the legislation did permit it. In his judgement, Chief Justice Burger stated:

> This is not to suggest that Section 101 has no limits or that it embraces every discovery. The laws of nature, physical phenomena and abstract ideas have been held not to be patentable. Thus, a new mineral discovered in the earth or a new plant found in the wild is not patentable subject matter. Likewise, Einstein could not patent his celebrated law that $E = mc^2$; nor could Newton have patented the law of gravity. Such discoveries are "manifestations of nature, free to all men and reserved exclusively to none."
>
> Judged in this light, respondent's micro-organism plainly qualifies as patentable subject matter. His claim is not a hitherto unknown natural phenomenon, but to a non-naturally occurring manufacture or composition of matter—a product of human ingenuity "having a distinctive name, character and use. . . ." His discovery is not nature's handiwork, but his own; accordingly it is patentable subject matter.

Chakrabarty's patent was subsequently granted, and other claims to the ownership of special microbial strains have been allowed in the United States. The judgement was clearly a signal to Congress that its legislative intentions had been explored by fine legal minds, and that if Congressional sentiment was that such patents should not be granted, then something must be done about it. The situation in Britain and Europe is not so clear cut. It is often said that organisms cannot be patented under British law, but a leading patent expert denied this in an article in *Nature* in 1980. When, some years ago, new yeast strains were patented in Britain, no one batted an eyelid. What the actual value of this ruling is to the biotechnology industry remains to be seen. Companies may rely on the practices of the micro-electronics industry in maintaining strict security, and going for rapid market penetration to safeguard their trading position. What you spare on patent attorneys, no doubt you spend on security systems.

All this takes us some way from the basic molecular and cell biology with which the chapter began. In a way though, this mirrors what has happened to research and researchers. More time is now spent with patent lawyers, venture capitalists and potential industrial sponsors, since the very practice of research in this area of the life sciences means that, fortunately or not, these questions are now likely to come up. Very unworldly and uncommercial scientists may well find their research has developed an industrial significance that was quite unexpected. Not everyone jumps for joy at that prospect, but only the most ascetic would ignore it and allow others to benefit.

Designer Genes

The experiments by Cohen, Boyer and their collaborators were an exploratory first step, intended to test whether the basic ideas would work in practice. They do. The next stage for those interested in the commercial angles was to try to get a human polypeptide made in a bacterial cell, by splicing the gene concerned into a bacterial plasmid. The group of scientists working on this, who were associated with the newly formed genetic engineering company Genentech, chose a small polypeptide, somatostatin, which controls the release of other hormones from the pituitary gland and is fourteen amino acids long. The actual DNA sequence of the somatostatin gene was unknown at this time so they designed a DNA sequence, using the genetic code, that would produce the correct amino acid sequence.

Having designed a gene, they then set about building it, first making a set of eight subunits, which were then joined together. Each end of the complete molecule was "sticky." Of its 52 base pairs, 42 ($=14 \times 3$) coded for somatostatin. The remaining ten were "sticky ends" and signals to the cellular machinery.

This molecule was then spliced into a bacterial plasmid called pBR322, to which had also been added the gene for the bacterial enzyme β-galactosidase, and the set of genes that control the synthesis of β-galactosidase, called the "*lac* operon." All this might seem complicated, but essentially the idea was to get the host cell receiving this enlarged plasmid to read the DNA sequence for β-galactosidase and then to keep on reading. The operon, a cellular control unit in bacteria, was there to switch on this operation. Several copies of the enlarged pBR322 plasmid were introduced in *E. coli,* and, as the bacteria grew, they made β-galactosidase molecules with an additional somatostatin tail. The hormone was then chopped off from the much larger bacterial enzyme and shown to be identical to that made in the pituitary gland. A common rapidly growing micro-organism had for the first time been persuaded to make a human hormone that it had never seen or heard of before. It may sound trivial, but it opened up the prospect of making a vast range of substances more cheaply in bacteria.

The technicalities of all this could be discussed at immense length. The point I want to bring out is the fact that right from the start genetic engineers began to design DNA sequences with particular principles or goals in view. In a sense, that is the skill that commands a high price, the ability to decide what kind of molecule to splice together, so that, when introduced into a host, the reprogrammed cell will immediately start to make large quantities of the molecule encoded by the new gene sequence. There are all kinds of technical tricks that make this process work. Gene-splicers know what string of genetic elements they need to edit together to make a "text" that reads well—i.e., one that bacteria, assuming they are the host organism, can read quickly to make a particular molecule. In some cases, the natural DNA sequence coding for a particular protein will be known, or can be quickly determined. There are ways of identifying the section of DNA where a particular gene is located and determining its sequence. Alternatively one can synthesise, somewhat laboriously, a DNA sequence that is equivalent to, but not necessarily identical with, the natural sequence. There are already gene machines that will make a small DNA sequence to order in a matter of hours.

It is worth bearing in mind that this feat of linking nucleotide units together to make a functional DNA molecule took hundreds of person-years in the 1960s, and won for two men, each leading large teams of researchers, Nobel prizes. It was then a decisive confirmation that the genetic code had been deciphered. It is now an element of the routine process of making molecules the biotech way. The most striking use of this technique was the construction by ICI scientists of a synthetic interferon gene (interferon is a substance that stops viruses infecting mammalian cells) 512 base pairs in length, or some twelve times longer than the somatostatin gene. Furthermore, they actually knew that their gene differed in a number of respects from the comparable one of the several interferon genes actually operating in human cells. They were, in a sense, experimenting to see if they could make different interferon molecules by modifying the design of the gene sequence. That is the destination of this technical virtuosity, to go beyond merely analysing how genes make specific proteins, to the stage of using an engineered nature to make molecules that improve on the natural design. I think this is stunning.

This confident approach to the *designing* of molecules has been around for a while. Already scientists in major drug companies are using sophisticated computer programmes to try to predict the shape and pharmacological activity of substances of a known composition. You take a chemical that you know something about and contemplate chopping off a small section or adding a new one. The question is, how the molecule will be rearranged by this chemical processing, how its three-dimensional geometry will change. Computer programmes now exist that can show, on a graphic display that simulates three dimensions, how simple molecules fold up. So, before you play around in the lab, actually making new molecules, you can look at their structure and manipulate them mentally before trying to synthesise them at the bench. One of the first university scientists to get involved with a genetic engineering company devoted to plant genetics, Martin Apple, of the International Plant Research Institute in California, is leaving to set up a new company that will specialise in the computer design of biological substances. Similarly a great deal of research now depends on access to enormous files of information, and biotechnology is no exception. Genetic engineers regularly consult computer archives of structural and sequence data. Without computers, then, it is virtually impossible to handle what is already known in biology, let alone play in one's imagination with possible new substances.

In Japan there is a man who is trying to build a machine that will seek to optimise polypeptide sequences for molecules with a particular function. Professor Wada's polypeptide writer is not intended to stop at mere design. When the optimal molecule has been "typed out," its manufacture is then to be transferred to a different biotechnology capable of mass production. Specially programmed bacteria will be set to churn out large quantities of the new molecule. Professor Wada calls this "printing for mass circulation," like putting haikus on cornflake packets. To do that one would need to put the genetic instructions necessary to make the molecule into host bacterial cells. The polypeptide typewriter is still in the experimental stage, and may turn out to be a beautiful idea, beyond the technology of its time, like Charles Babbage's Analytical Engine, a calculating machine designed in the early nineteenth century but never built on its intended scale.

Scaling Up: Industrial Microbiology and Cell Culture

The first gene-splicing experiments were done on the laboratory bench, and the amount of protein churned out by the reprogrammed bacteria was incredibly small. To translate these ideas into an industrial process, capable of generating pounds or tons of material, requires a massive increase in scale. The yield of somatostatin from the procedures just described was around 10,000 somatostatin molecules per bacterial cell, which was encouraging. The next series of experiments, that made insulin in the same way, were more productive and yielded 100,000 molecules per bacterial cell. This was sufficient to produce an eventual yield of 100 grams of insulin from a 2000 litre fermentation vessel. Insulin is normally extracted from animal pancreatic glands. To make 100 grams that way takes about 1600 pounds of material from the slaughterhouse.

But even a 2000 litre fermenter, which is a small one by industrial standards, is a great deal larger than a culture dish or a lab-bench fermentation vessel. With that change of *dimensions* comes a whole set of engineering, biochemical and economic problems that do not exist with bench experiments. The trick is to keep billions and billions of micro-organisms in the conditions they like best. If their nutritional requirements can be met, and if they can be prevented from poisoning themselves with their own waste products, then they will grow like crazy. Moreover, bacteria can be persuaded to devote an extraordinary proportion of their cellular resources to the synthesis of one particular chemical. The result of this dedication is that they are incredibly productive as microfactories. Their output can be prodigious, as long as the growth conditions remain stable. If they do not, then the bugs soon die, in their trillions, with nothing useful to show for it.

So fermentation on an industrial scale, long practised in brewing and the food industry, tried and then abandoned in the chemical industry, and developed for pharmaceutical production since the Second World War, requires a number of skills—in plant design, in control engineering and in understanding the dynamics of fermentation under production conditions. Essentially, the task is to provide a carbon source, like sugar or starch or cellulose, a nitrogen source, such as ammonia or atmospheric nitrogen, and mineral salts containing phosphates and to keep it all mixed. In most cases, the incoming media have to be sterilised to kill off contaminating micro-organisms that would compete with the bugs that one is trying to grow. That can be difficult to achieve efficiently. In the simplest systems there is a paddle in the centre of a cylindrical fermentation vessel which turns, swishing the culture around. Larger, more advanced systems may use different methods of agitation, such as an air jet and more complex patterns of circulation.

Then there is a whole range of biological problems with the culture of bacteria. One possibility is that they start shedding genes to make life easier. For example, a bacterium with a plasmid full of insulin genes is operating under a handicap, compared with more prodigal relatives that have dumped their plasmids. If the habit spreads, then you end up with a lot of bacteria and precious little insulin. One way of stopping this happening is to pick a plasmid conferring antibiotic resistance and then to dose the fermenter with that antibiotic. Any bug getting rid of its plasmid gets wiped out by the antibiotic. But that is expensive. Another problem is marauding bacterial viruses that delight in busting up bacteria. The bigger the fermentation vessel, the more material you lose if things go wrong.

Having grown up a large volume of micro-organisms, either in a batch or in a continuous-flow process, one then has the task of extracting the substance of interest from the resulting thin bacterial or fungal soup. It is actually possible for fermentation to occur in the solid state —any compost heap bears witness to that. Similarly, it is how the Japanese make tofu, fermented bean curd. But most industrial fermentation occurs in a liquid medium. If the product of interest is whole cells, as is the case with single cell protein, then you just have to strain off the liquid and dry the cellular material, which tends to use up a great deal of energy. If the product is a liquid, like an alcohol, then you have to separate it from any troublesome fermentation by-products, for example by distillation. If the product is a molecule, then fishing it out may be a problem, particularly if the bacteria have accumulated it inside themselves and not secreted it into the surrounding medium. One possibility is to use cells with "leaky" walls that let the material out. Even then you have to extract that molecule from the culture medium. With medical products very high levels of purity are essential. Johnson and Johnson, the drug company, has even booked space on the space-shuttle to try purification by electrophoresis in the gravity-free conditions of outer space.

So far I have been assuming that the organisms in the fermenter are bacteria, yeasts or moulds. But, as I mentioned in chapter 1, one can grow plant cells in culture vessels. It is also possible to grow mammalian and human cells in these conditions. Technically it is tricky, because they are more fastidious in their food requirements and they tend to clump together. But interferon can be produced from large cultures of fibroblasts—immature cells, obtained in this case from human foreskins.

It is possible in some cases to dispense with cells altogether and promote a particular reaction with millions of enzyme molecules bound to a ceramic, plastic or organic support. These are called immobilised enzymes. Recently it has proved possible to immobilise cells, which means that their metabolism is switched off and they act as an inert matrix for a particular enzyme locked within. The use of these procedures is expanding, but in most cases a fully functioning cell is the necessary context for and site of biotechnological processing.

The business of keeping billions and billions of these specialised organisms alive is actually very complex and demanding, although it tends to be downplayed all the time, compared with the tasks of the superstars who create them as specific strains or cell lines. This, then, is production experience that few university scientists have, and the genetic engineering companies, as they move into the production phase and turn to new host organisms like yeast, are recruiting industrial personnel with knowledge of fermentation from brewing and the drug industry.

The skills of applied microbiology have their roots in traditional industries like brewing and cheese-making. In the late nineteenth century, the new sciences of microbiology and biochemistry were recruited to throw more light on what actually happens in industrial processes. But the new disciplines of fermentation chemistry and industrial microbiology remained low on the academic totem pole, and their practitioners enjoyed no great status. The importance of fermentation to the postwar pharmaceutical industry did nothing to change that. Certainly molecular biologists thought their prosaic, but necessary, investigations were unutterably boring and trivial.

As biotechnology enters a new phase and "scales up" its brilliant insights into industrial processes with marketable products, this social order may very well be challenged. It is a truism of the traditional chemical industry that more than half of your costs go on fishing the desired product out of the muck at the end of the chemical reaction. This is "downstream processing," and the engineer who can get it right—sometimes by the most mundane adjustments—is the person who can make the difference between profit and loss. In the same way that molecular biologists were disdainful of mere biochemists, their progeny, the gene-splicers, convey an air that they can be the mandarins of a new chemical industry while remaining above the grit, grime and smells of chemical factories. They are beginning to learn that bluff practical engineers with dirty fingernails may have at least as much to offer as the new biotech millionaires with gold jewellery and designer jeans.

4

Selling a Medical Future: Products for Profitable Diseases

In this society, our life needs are largely addressed through what can be manufactured and sold to satisfy them. This means that such needs are met in a particular way. Ideas about what is desirable are screened for those that are saleable—either as goods or as services. Public interest in what is being produced for us has little if any impact on the development of such commodities. If public concern has any role, it tends to come when products have already been launched into the public domain. And, with today's techniques of advertising and marketing, a strategy for handling such attitudes will already have been prepared. By then people's feelings will have been measured in advance, their choices anticipated and expressions of need deftly redefined and tailored so that they can be satisfied by what is to come on to the market.

In the field of health care, something like this is being practised all the time. The aim is to make money out of the desire for health, through the massive consumption of goods and services. Two thirds of the world's population are carefully serviced by a powerful industry, based on pharmaceutical products. The remainder, the desparately poor, are excluded. To pose the main question that this chapter asks, it has to be put in a specific order: How will this industry use biotechnology so that it fits into its commercial plan for health?

At this point, I should say that many of the medical applications of biotechnology will reduce human suffering, will allow some diseases to be treated more effectively and illuminate areas of present medical ignorance. That counts as progress. But I am not persuaded that they will bring down the cost of medicines, or focus more attention on the social causes of disease, or give people greater control over the way health care is developed and delivered, either to the community or to themselves personally.

Corporations which supply the health industry with their drugs occupy one side of a threefold relationship of power. They supply a medical profession geared to "acute," curative treatment (treating a patient who is endangered by a disease already contracted), and this practice is heavily oriented towards the use of those drugs. The companies need to secure this as a dependable market. The dependency, therefore, is aided and abetted by vigorous advertising and marketing campaigns directed at doctors by such companies. The last side in this triangle is filled by those who are medically defined as "patients"— people who are thought to be in need of the help that the other two can supply. They appear in this context because they have become ill. In this state they are passive; their health will be administered to them. Power remains with the other two. This relationship is being reinforced by the health industry's use of biotechnology. This is why my initial question was posed in the way that it was. Such a result would be a shame, and something to be weighed against the progressive aspects of the new technology. It is the feeling that we need such an evaluation, a public weighing of the gains and losses, that impels me to write; of course, it is just this need that the advertising rhetoric of biotechnology denies.

A basic aim of my argument is that we should react critically to the rhetoric that already surrounds biotechnology. Everywhere you look you find hyperbole and grotesque simplification. This product will cure cancer, this technique will solve the world's food problems, and this idea will make drugs cheaper. To get ideas from the laboratory bench into the market place takes sponsorship, enthusiasm and *promotion*. At every stage in that process selling is going on. Research scientists in industry lobby their director, he cajoles the main company board; they talk to their bankers or to industrial partners, or to the government. Independent entrepreneurs hawk their wares around the finance houses or the venture capital concerns. Scientists campaign for more money for their speciality. Every so often a decision is taken to use a new process or to market a new product and then the advertising begins, with only the consumers and their regulating guardians left to convince.

There has been much speculation about what will be the "first" product to emerge from the biotechnology companies. After all, some of the newest operations like Genentech or Cetus are now substantial businesses, capitalised at several hundred million dollars with a turnover of perhaps $5–10 million. At the moment their money comes from services, not from goods. They sell laboratory skills not products. Others are extensions of existing businesses, like Bethesda Research Labs, which sell speciality chemicals to researchers, or Novo Industri, which sells enzymes for detergents amongst other things.

Until mid 1982, no product derived from gene-splicing had been brought on to the market. At one point it looked as if insulin might be the first. However, the honour goes instead to a vaccine for pigs and calves to prevent them from getting dehydrated by diarrhoea and departing this life before they can be profitably sliced into bacon or veal. This product comes from Intervet, a subsidiary of the Dutch chemical multinational AKZO, which is one of the five biggest companies in Holland. One up to the Dutch.

Vaccines are just one avenue of research; we come back to them in a while. The list of medically useful molecules that bacteria and other cells are going to be encouraged to knock out in thousands per second is already long and growing every week. Insulin and interferon are well known, and we turn to them first. We can add a wide range of hormones like somatostatin, somatotropin, which is a growth hormone, interleukin, calcitonin, relaxin, cortisone, gastrin and thymosin, which orchestrates the immune response. Then there is the myriad of proteins in the blood, like Clotting Factor VIII, which most haemophiliacs lack, and serum albumin. Then we can add the variations on the forms found in nature, hybrid molecules that may evoke a more powerful response than the "natural" forms, and proteins bound to other molecular complexes, in order to reach particular tissues in safety.

All these represent a small part of an explosion of possibilities, a galaxy of potential products for the pharmaceutical companies and those corporations who are starting to join them in the health business from the food and chemical sectors. It will certainly increase the repertoire of the most highly trained experts in our culture, the doctors, but will it make us healthier? That depends on your view of disease and what medicine does to counter it. The question of just which medicines are likely to get developed is therefore central to this issue. After all, there are already many safe and efficacious drugs that languish without a manufacturer willing to make and sell them, because the demand is thought insignificant. If organisms, from bacteria to plants, are screened to find the most profitable, so too will the health industry screen for the most profitable diseases. Moreover, curative medicine is important; we all need it often and will continue to do so.

But preventive medicine, in the form of vaccination, public health programmes, health education and occupational medicine for a safe working environment, is at least as important. Biotechnology looks as if it will be used predominantly to develop curative medicine. It will be a source of products, technical fixes for some diseases, the use of which will leave intact the social and economic processes that make us sick.

Insulin: Out of the Abattoir into the Lab

Word is that it was the success of the Genentech team, in 1977, in getting the gene for human insulin to operate in a bacterium, that really convinced molecular biologists that recombinant DNA research had commercial potential. Before then, so the story goes, most scientists were impressed only by its technical value to their research. Afterwards they began to make friends with patent lawyers and to be propositioned by bankers.

The actual historical situation was obviously more complex than that. Certainly some companies like Schering and ICI had already weighed up recombinant genetics in the early 1970s, when it first appeared, and decided that it would soon be useful to them. Some scientists, like Boyer, were also very quick off the mark into commercial research, with some urging from entrepreneurs like Robert Swanson, the president of Genentech. But it took a product like insulin, well known, widely used, almost a household molecule, to make the point to the majority.

Insulin is something on which millions of diabetics are continually dependent for their health and well-being. It controls the body's storage of sugar. Without it many diabetics would die; with it, they can usually lead relatively normal lives. It is in effect a bulk chemical, and the market in the United States alone is worth $200 million per year. Researchers realised in 1977 that if Boyer's team had already got bacteria to make insulin, a substance that no bacterium had ever made in nature, then maybe the prospect of making big money out of one's research skill was not a hopeless fantasy. Insulin seems an obvious choice as a commercial goal for this technology since it meets so many of the criteria for a successful, profitable product.

First of all, insulin forms the basis of an *established* market, and one that is still growing. To put it another way, for sixty years it has been accepted that the drastic effects of diabetes on life and health can be held in check by regular injections of insulin. The need is then clear cut, and clinicians need no persuading to use it as a form of therapy. Secondly, it is a *mass* market. In the US, it is dominated by one powerful producer, the Eli Lilly Company, and, together with another, Novo Industri of Denmark, it has joint control over eighty percent of the world market, worth some $400 million in 1981. Even a small share of this market is worth having then. Thirdly, this market is expected to double by 1986, so rapidly is the incidence and diagnosis of insulin-dependent diabetes increasing. Some forecasts show that demand will soon outstrip supply. The rate of increase of diabetes, both within particular countries and world-wide, is such that the extraction of insulin from the pancreases of slaughtered cattle and pigs, the present source of supply, will be insufficient to service the market at some point in the not too distant future.

Furthermore, because beef and pork insulin are not chemically identical to human insulin, a significant percentage of diabetics start to produce antibodies to their injected insulin, treating it as a foreign protein. Their bodies, in other words, start to destroy the insulin before it can be used. If one could make an insulin that was a perfect copy of the human molecule then these problems should not arise. All these considerations add up to an enormous incentive to develop recombinant bacteria that can synthesise human insulin.

By 1981, the substance was at the clinical trials stage, with human volunteers receiving shots of it. By 1982, advance publicity material was being sent out by Eli Lilly, in anticipation of its eventual marketing. In May 1982, they advertised for some new sales staff. In July 1982, the company flew about forty European journalists to San Francisco for five days, as part of the build-up to the eventual launch. All this was part of a massive promotional effort.

Since the 1950s that has been the key tactic of the pharmaceutical companies—relentless promotion of particular products to capture a large market share and recoup the very high development costs in the remaining lifespan of the patents protecting product or process.

However, in this case, in 1980, Novo Industri announced that it had developed a process for converting porcine insulin into human insulin, by chopping off the final amino acid of one chain and adding the one that is there in the human hormone. Quite why this had taken so long to achieve I don't know. It is said to be a difficult operation to scale up to an industrial level of production. It was a clear challenge to the bacterial process from a company that had been marketing insulin since the 1920s, and which had a substantial market share in Europe.

Novo gained a licence to sell its new insulin in Britain in June 1981. Despite this, and perhaps because of the anticipated shortfall in the supply of materials from the abattoirs of the world, the plans of a number of companies to market a "bacterial" insulin are still going ahead.

In September 1982, an article appeared in the *Guardian* which presented the competition over how insulin was produced as a titanic struggle between Eli Lilly and Novo Industri, each backing a different process, each with its reputation on the line. Both insulins are now on the market, along with the range of existing products. It remains to be seen whether there are significant clinical drawbacks or advantages to either of them, or, indeed, whether on balance any major net advantages will emerge. The price of the new Novo insulin is around £3 a dose, forty pence dearer than the price of its porcine insulin. Enquiries to Eli Lilly revealed a spread of figures for the cost of a dose, depending on the formulation, though a report in *Nature* in September 1982 stated that Eli Lilly wanted to undercut the Novo price by twenty per cent. Whether these prices can be used for meaningful comparison at present I doubt, because very complex marketing must be going on.

Interestingly, marketing like this is a mode of operation that is outstandingly successful commercially, with the profitability of the big pharmaceutical concerns considerably above the industrial mean. On the other hand it is not so clear that the public have actually benefited from these business practices. Markets for specific drugs tend to be shared out by two or three large multinational concerns, which are in a position to control prices and to conceal the level of profit from governments seeking to investigate why they are being asked to pay such massive and rising bills for drugs. On occasions, drug companies have been forced into returning a proportion of profits to the consumer, a clear admission that overpricing has occurred. A possible defence of these now well developed practices, apart from the commercial argument that risk-taking in business must have adequate rewards, is that they result in a steady flow of new products, a few of which at least do mark a real advance over existing medicines. The price of new drugs may be high, but, say the defenders of the pharmaceutical companies, it is a price that is worth paying. The crucial assumption in this argument is that improvements in health must follow, and can only follow from the appearance of new drugs—or from improvements in clinical medicine.

The case of diabetes provides a powerful illustration that this argument is not valid, although that does not in fact make the resulting political decisions any easier. When you look at a graph of the incidence of diabetes against time, an obvious question to ask is why the condition should become more common. There is persuasive evidence to suggest that although a tendency to manifest the symptoms of diabetes may be genetic, dietary factors play a large part in determining whether such potentialities will be realised. Put crudely, most diabetics are made by their diet, even if potential diabetics are born. Given the levels of sugar, fat and low-fibre carbohydrate consumption that constitutes the Western diet, many people will develop some of the symptoms of diabetes in their lives. On a high-fibre diet *far fewer of them* would do so.

This model of the cause of diabetes, which is not proven but is certainly persuasive, is based on studies of people drawn in to Western culture by economic developments and urbanisation, whose diets and patterns of disease can be shown to have changed as a result. In Kenya and Uganda medical records exist that span the period of urbanisation that brought landless Africans into the towns. The changes in their diet can be documented, and with them has come an increase in heart disease, cerebrovascular disease, hypertension, obesity, diabetes and a range of other conditions, called by Burkitt and Trowell "Western diseases." Similar patterns can be found in the Pacific, in Asia, among American Indians and in North Africa and the Near East.

Let us assume for a moment that this model of the causation of diabetes is proven, which as yet it is not. The clear implication is that to improve health standards people should eat less sugar, salt, animal fat and less highly processed, fibre-depleted foods. The incidence of diabetes would then go down, just as that of heart disease has done. If somehow the dietary trends of the last two hundred years in the industrialised countries, or of a much shorter period in developing countries, could be reversed, then the prevalence of diabetes would fall and with it the world demand for insulin. Does the prospect mean that attempts to produce insulin from reprogrammed bacteria are unnecessary? I don't think so.

Diabetes is a common condition. In a country like Britain there are 600,000 diabetics, many of whom need insulin daily. That is one in ninety people. Even if dietary patterns could be changed, many of these people would continue to need insulin. Moreover, the rate of incidence of new cases would not fall to zero even if the shift in eating habits was massive and sustained, which it would not be. Many thousands of new patients then would require insulin in future, even though the overall level of demand would fall. While it doesn't seem likely that in the longer term the new insulins will be offered at a cheaper price, they may offer clinical advantages. If "local" companies can make them, assuming patent protection and the buying up of licences does not confer control on dominant concerns like Eli Lilly, then industrial countries like the UK could save foreign exchange by not having to import drugs that doctors may wish to prescribe.

But there is a clear need to try and reduce the demand for insulin, which the advertising rhetoric of cheaper, better insulins obscures by suggesting there is no alternative—disease is fixed, it is always with us so we can only cure it. Clearly there are alternatives, but it means acting against the effects of urbanisation in developing countries, against the promotional activities of the food processing industries and the carefully nurtured preferences of busy, anxious people for too much sweet, fat-laden, salty and soft foods. But if we don't try to do that, hard as it will be, then more and more people around the world will suffer the effects of diabetes, which are not held in check by insulin injections. What many nondiabetics and some diabetics fail to realise is that even if one's body can be persuaded with insulin injections to deal with sugar in an effective way, diabetes still often brings with it damage to the eyes, to the heart, to the circulatory system and the extremities of the body. For example, if the incidence of diabetes increases, more people will contract gangrene. That is not a situation that the drug companies create, but it should be stressed that they are complicit in its continuation by concentrating their research and marketing efforts in the directions they have.

When biotechnology is used in this way, it becomes a powerful tool in the development of corporate capital. But this is not a reason to condemn it out of hand. The very great positive advantages it might bring depend upon what choices are made about its directions and use. It is an argument, then, about the social relations which set the priorities for, and get embodied in, science—health care in this particular case. We cannot separate the need for a public debate over biotechnology from all the questions that these social relations raise. Above all, this means weaning it from the hype that surrounds it and its applications. The story of interferon is another demonstration of this need.

Interferon: The Rise of a Fallen Molecule

In 1957, a British virologist, Alick Isaacs, and his Swiss collaborator, Jean Lindemann, showed that cells from chick embryos could be protected against viral attack by some substance secreted by cells already challenged by viruses. It was as if one element in the cells' defence was secreting a substance which interfered with further infection. Isaacs called this substance interferon although others called it "misinterpre-ton." It proved to be extremely active in tiny amounts and very difficult to purify. It excited immediate interest as a possible way of treating viral diseases like measles, flu, herpes and smallpox. Hopes were gradually built up that here was the viral equivalent to the antibiotic penicillin. It would be a "wonder drug" that would stop viruses in their tracks, if it could be made in quantities sufficient for clinical use.

A committee was set up with representatives from both British government funding and sponsoring bodies and industry to look into the possibilities of exploiting this research. This was stimulated by commercial inquiries from an American firm and continuing embarrassment in the UK that the patents on penicillin had been passed to the US. The key to success was the ease and economy with which interferon might be made in industrial quantities. The technical difficulties proved insuperable, and industrial interest in interferon fell away in the mid 1960s. Isaacs, himself a somewhat tragic figure, who battled to keep his vision for interferon alive, died at an early age in 1967.

A few investigators ploughed on relentlessly, trying to find ways of increasing the yields of interferon from the various systems in existence. A Finnish scientist, Kari Cantell, who was able to use the unwanted white blood cells from the Finnish Red Cross blood donation programme, became in effect the world source of supply. Even today, his laboratory is the major supplier. However, the difficulty of extracting interferon means that the price is astronomical. For example, in 1978, Dr. Cantell's laboratory processed 90,000 pints of blood and produced only one tenth of a gram of pure interferon. (But even such a tiny amount would be enough to treat 200 patients with chronic viral diseases.)

A new situation appeared in the mid 1970s. Interest grew dramatically in the possibility of using interferon for the treatment of various kinds of cancer. This arose partly through the efforts of Mary Lasker, the influential advocate of increased expenditure on medical research, in orienting the American Cancer Society to the use of interferon. Although the process by which particular cells change their state and start to form a tumour is reasonably well understood in some cases, nobody really knows what sets off this process. We do know that a class of substances called carcinogens—which includes a vast number of such things as asbestos, vinyl chloride monomer or benzene—can start the process going. We also know that in mammals and birds some viruses can have this effect, but until very recently no one had shown satisfactorily the existence of viruses that will make human cells cancerous. Then again, we also know that a predisposition to develop certain forms of cancer can be genetic. In other words, the cause of cancer seems likely to be a complex process that involves some or all of carcinogens, viruses and genes. One implication of this is that a substance that inhibits viral infection might possibly be a powerful anticancer agent also.

At the same time as it was arousing this kind of interest, new methods of producing human interferon from cell cultures appeared. Then, as the biotechnology companies were set up in the late 1970s, their interest turned to the production of interferon in bacteria. In January 1980, Dr. Charles Weissman, of the University of Zurich and Biogen SA of Geneva, announced at a press conference in New York that his team had succeeded in cloning human interferon genes in *E. coli*. This incident, mentioned earlier in chapter 2, drew protests from scientists, who felt that untested assertions were being made purely for publicity and promotional reasons. But interferon is surrounded by hype. Potentially, it is the most glamorous product in the present biotechnology stable. To produce, test and license marketable quantities of interferon is one of the biggest prizes for biotech companies. Since Weissman's announcement, the pace of research and development has become dramatic. Genentech, Biogen, Cetus, Genex, Hoffman La Roche, G. D. Searle and other concerns are all working hard to get an interferon product into the market.

However, the results from the clinical trials are mixed and the mode of action of the interferons still unknown. What this intense burst of activity has made clear is that the interferons are a family of molecules, some twenty at least, produced by a number of genes. This finding has prompted the construction of "hybrid" interferons, artificial variants on molecules specified by the set of recently discovered genes. This deliberate shuffling of the molecular pack permits the search for more effective antiviral or antitumour agents, but it also suggests a route to obtaining patents on molecular "products of manufacture." The case for patenting natural interferons would be weak; that for patenting "new" molecular "products of manufacture" might be stronger.

What are we to make of all this activity? Interferon has suddenly become a glamour substance, an elixir, a possible wonder drug again. There are sad stories of cancer patients paying enormous sums for the low-grade black-market interferon in the vain hope that it might help them. There are reports of physicians setting up funds to collect money for their patients' interferon. In the United States the Food and Drug Administration has recently condemned the sale of an interferon preparation as a food supplement, like vitamin C tablets. In the Soviet Union, interferon is sold in extremely low, probably useless, doses as a nasal spray to cure colds. Interferon, then, stirs the hopes of many people. Is this the kind of goal that should be selected for medical biotechnology? As with insulin, there is no clear-cut answer, and evaluation of what has happened depends on the view taken of the causes of disease and ideas about health priorities. If it is believed that medical research should be organised primarily around research-intensive hunches devoted to uncovering therapeutic substances for use in high-technology, curative medicine, then the interferon drama, a sudden upsurge of "heroic" research in the hope of finding a magic bullet for the dread disease, must seem rational and defensible. Even though the clinical trials have now shown that interferon is not an exceptional chemotherapeutic agent for cancers, more is now known about the interferon system, and the possibility of its use for a wide range of serious viral diseases is still appealing.

On the other hand, some people think that biomedical research should be oriented much more strongly to prevention, particularly with a disease like cancer, where a significant proportion of cases come about through environmental factors, such as continual exposure to carcinogens in the workplace. On this view the sudden channelling of resources into interferon production, to establish a new mass market, is a depressing extravagance, a lurch towards profits at the expense of health standards.

I have of course presented these views of medicine and medical research, as mutually exclusive, as though one could *either* have prevention *or* cure. That is not true, and it would be wrong to imply *that in any simple or direct way* effort or investment in interferon research subtracts resources from the prevention of cancer. *If* such an effect occurs, then it occurs through much more complicated processes. These same considerations apply to insulin and diabetes. Certainly the very public, carefully managed excitement about the bacterial synthesis of interferon, which had a rather obvious financial dimension to it, can only have helped to strengthen the deeply misleading view that a substance like interferon could in itself be an answer to the cancer problem. That is a shame, because there are all kinds of less spectacular ways in which we could seek to prevent cancer. But they don't fit as neatly as interferon does into the socio-economic structure of the contemporary industrialised nations.

Similarly to see interferon as *the* answer to the burden of viral disease is grotesque, because millions of people stricken with measles, polio or hepatitis are likely to remain too poor or too deprived of medical care to get interferon, and too malnourished to be able to resist the effects of a disease like measles or influenza. Many viruses only kill those already weakened by poverty and exploitation.

Growth Hormones: Spot the Market

So far we have been considering substances for which the probable market is very large. Indeed one of the issues I have been discussing is why—socially and economically—the market or the demand is so large and whether it need continue to be so. In the case of human growth hormone the market is nothing like as great. Perhaps 1 child in 5000 suffers from growth retardation due to growth hormone deficiency, and in Britain twenty-one clinics administer the hormone to around 600 children a year. In America it is thought that 2000 children a year receive injections of growth hormone three times a week. This in itself is no mass market, and even allowing for a substantial profit out of each patient's dose of hormone, no company could earn very much from this area as things stand.

At present the hormone for one child's treatment costs around £10,000 per year. If new bacterially produced hormone could be marketed at a lower cost than this, but still allowing, say, £5000 profit per child, which would still be a saving to the National Health Service, the total revenue would be around £3 million per year. That is such a small sum for a big company that it would only be appealing if the product could be developed so as to take in a bigger market. So why is growth hormone up among those products that the genetic engineering companies are trying to get on to the market?

Growth hormone therapy has its roots in clinical research. The production of the right amount of growth hormone for the right period of time is vital to normal growth. The hormone is secreted by the pituitary gland at the base of the brain, and its role is to orchestrate the bodily processes of growth. If none is produced then dwarfism results.

It is possible to extract the hormone from preserved pituitary glands taken from dead bodies. This indeed is how growth hormone has been made available to clinics in the UK from the early 1960s, in a scheme funded by the Medical Research Council (MRC). Under that scheme, at the peak of production, 42,000 glands were extracted every year, which was sufficient to meet the national demand. On that basis, about 70 glands were necessary to provide the hormone for one patient for one year. The reason that there are now fewer people with seriously retarded growth than thirty years ago, is successful treatment.

In 1977 the administration of the scheme was handed over to the Department of Health and Social Security, the government department in overall charge of the National Health Service, and the production of hormone has since fallen. There are said to be several reasons for this. Firstly, there is a possibility that the legal basis on which organs were removed in hospital and public mortuaries was unsound. Certainly one regional health authority stopped the practice in 1981, and press reports suggest that the NHS management were surprised that this was going on. Secondly, a change in the method of payment to mortuary attendants for carrying out this unpleasant task is said to have decreased the supply. Thirdly, there is the clear implication in the statements of some of the people involved, that the administrative arrangements were simply not so efficient. Production has now been transferred to the Centre for Applied Microbiological Research at Porton Down, once a germ warfare establishment, where work is also going on to produce growth hormone from bacteria. The alternative method is to transfer the genes specifying the hormone into a bug like *E. coli.* and to extract it from the bacterial culture. This was first done in 1979. The expectation is that this mode of production will be substantially cheaper than that from the mortuary route, with the consequent saving to the NHS.

Here then is a substance that is urgently needed by a few people. The present way of producing it is expensive, unpleasant and open to abuse. Demand in the UK at least apparently roughly matches the supply. It may well increase slightly as more cases of treatable growth abnormalities are recognized. World-wide the situation is different. In 1982 the president of Kabigen AB, the genetic engineering company set up by the Swedish concern Kabi Virtrum, suggested that only one sixth of all cases of hypopituitary dwarfism can be treated, given the available supply of cadaver material. He also stressed that despite meticulous purification and testing procedures one cannot rule out the possibility that some growth hormone produced in this way will contain so-called "slow viruses," which infect brain tissue. Bacterial synthesis removes all these problems.

It has been suggested, however, that cheapening growth hormone might lead to its abuse by people wanting to be tall but who are not pathologically short. Knowing that height is a source of anxiety to many people, particularly adolescents, it seems possible that unscrupulous suppliers might seek to market height-augmenting drugs.

This is certainly possible, but it seems to present no more and no less of a potential problem than the abuse of anabolic steroids in sport or benzedrine for pleasure. It is an argument for regulation, which is required in any case, and for medical control over the supply, which would also occur. It does not invalidate the project of producing growth hormone in bacteria.

But the question remains why it should be of interest to companies orientated to mass markets. The historical record clearly shows that drugs for which demand can only be small are "orphaned" by commercial processes of selecting possible products. Certainly the estimates of the market for growth hormone imply that other uses are foreseen. In 1982, the business manager of Celltech in Slough stated that the US market was around $100 million per year, which is many times what is needed to treat hypopituitary dwarfism. One possible answer is that growth hormone will be used in a novel way to accelerate tissue growth and wound healing after surgery, to help fractures mend and to assist the treatment of burns and ulcers. Another answer is that hormones from all mammalian species are chemically fairly similar. So learning how to make human growth hormones is just like learning to get the growth hormone for cattle or pigs or sheep from bacteria, and the markets for these substances is enormous. One estimate in 1982 was that $500 million worth of beef and pork growth hormone could be sold per year, five times the market for the human hormone. Indeed, growth-promoting substances are coming to be used on a vast scale in agriculture, since they decrease the time and money needed to fatten livestock to a given weight for sale.

The problem, however, is that this practice leaves hormone residues in the meat for a certain period after slaughter. If the necessary time is not allowed to pass between slaughter and consumption, then people eating that meat will receive harmful doses of growth hormone, which can have serious medical side-effects. There have recently been reports from Puerto Rico of endocrine problems amongst children caused by illegally sold hormone-bearing meat. In the summer of 1982 French agricultural authorities announced that lamb bearing unacceptably high levels of hormone would be banned from imports, partly to anticipate similar medical problems. Just as with antibiotics, hormones have a use in the clinic, but when used in agriculture in an unregulated way they may create significant health problems.

Blood Products: The Struggle for Private Ownership

Blood is a cocktail with hundreds of ingredients. Basically, they fall into two main groups, the circulating cells like the erythrocytes, which transport oxygen, and the fluid mosaic of proteins that make up the plasma. The most abundant is the protein, serum albumin, which maintains the volume of the circulating blood. Other serum proteins include the elements of the coagulation system which are concerned with the formation of a blood clot at a hole in a blood vessel wall. Clotting Factor VIII, the protein that most haemophiliacs lack, is one of these.

Plasma can be easily separated from the whole blood. It is more difficult, but now possible, to isolate specific fractions from it, using technology adapted from the dairy industry for the separation of curds from whey. The Second World War produced a lot of work on plasma substitutes for use in the battlefield, and in 1946 a team at Harvard described the physicochemical procedures for isolating particular proteins. That work and related projects on haemoglobin provided a lot of fundamental data about protein structure upon which molecular biologists drew after the war.

In Japan, too, similar studies had been going on, but under conditions of indescribable brutality at a prison camp at Harbin in Japanese-occupied Manchuria, where prisoners of war were used in live experiments. When the war was over the Japanese officers and scientists running the camp did a deal with American intelligence officials, exchanging data from a wide range of experiments in return for their freedom. Just how they were in a position to suggest such an arrangement, or why the Americans accepted it, has not been made clear. One of the scientists from the camp joined a pharmaceutical company, which soon after began to market the first artificial blood plasma.

The technology of plasma processing has advanced considerably since then. It is now possible to pass a patient's blood through a bed-side blood-processor that removes specific celltypes or blood fractions, such as antibodies. It is also possible to remove donated plasma from individuals who have especially potent or rare plasma constituents. Donations of what are very large amounts of protein are a risky procedure. In America up to one litre of plasma per week may be taken from donors. In Europe the recommendation is that only a quarter of that be removed per week. The losses in the donor are made up with serum albumin itself derived from other people. Alternatively, the plasma can be processed at a central facility, having been separated from whole blood at regional transfusion centres, and fractionated into the individual protein constitutents, such as the Factor VIII required for the treatment of haemophilia.

In the United States, Genentech announced in 1982 that it had got human serum albumin expressed in *E. coli* (that is, the bacteria were making the protein, probably in very small quantities). The prospect is there of making massive amounts of serum albumin in this way, possibly in other bacteria or using yeast as the host organism. This work was done under contract to Mitsubishi, the Japanese conglomerate, which retains the exclusive marketing rights. The estimated world market is around 100 tons per year, which would bring sales of $500 million. In effect that makes it a bulk chemical. One can be reasonably sure that other companies have selected this goal as well, with less publicity. It seems to be the Genentech style to ensure that each scientific success and its commercial basis gets wide media coverage. Other companies go to considerable lengths to avoid just that.

Blood products are an area of growing competition in biotechnology, and with continuing innovation our attitudes to and use of blood will change. Whilst there may be clear gains as new products become available, and as some of the health hazards in existing donation programs are removed, it is also worth asking at what price and to whom will these benefits be available. In Britain, the struggle is to keep blood donation and processing in private ownership. In America the issue is one of who will control the standards and prices of the private supplies of the new blood products.

Monoclonal Antibodies

In the previous chapter we considered a case where researchers and their sponsors pulled together to get patent protection for their work. Not only did they have to act decisively in the early days; they had to keep up the pressure, shrugging off criticism and abuse. The case of monoclonal antibodies makes a complementary story, for in this case what Cohen and Boyer must have feared actually happened. Someone else got the patents on the research, through being swifter off the mark and being prepared to face the outrage. As with recombinant

DNA research a series of experiments have developed into a research procedure of major significance and the basis of a new branch of industry, conducted by small university-linked companies and major concerns buying into the centres of excellence.

Higher organisms have evolved a sophisticated and flexible defence called the immune system. This is located in the veins and arteries and in the vessels of the lymphatic system. Various sets of cells in the body are organised to detect and destroy foreign substances, like bacteria, viruses and extraneous molecules, such as those of pollen, that pose a threat in some way. As such substances, called "antigens," enter the body, special cells churn out large numbers of complex molecules, the antibodies. These have structures that enable them to lock on to particular antigens. This locking is very specific. Particular antibodies will fit only one kind of antigen. Foreign substances are thus blocked in their action and can be more easily destroyed by a further wave of defending cells.

This process is the basis of vaccination, to which we shall turn in the next section when we consider how recombinant genetics can be applied to the creation of new vaccines. A vaccine is a suspension of inactivated or killed viruses or bacteria, that on injection stimulates the production of antibodies without causing disease. As a result, the vaccinated person or animal acquires an immunity against that particular disease, since a stronger immune response occurs on the next occasion that the disease agent appears. Similarly, sera and antitoxins can be created by exposing to infection another less susceptible organism, say a horse or a rabbit, and drawing off the resulting antibodies from its blood, which can then be injected into someone at risk from the disease concerned. Most substances have a number of antigens on their surface: that is to say, they are perceived by the immune system as highly complex structures requiring a range of specific places to attack. Consequently, the immune response consists of the production of a mixture of several or many antibodies, so that the biological picture is rather confusing and difficult to interpret.

The immune response has been fruitfully explored at the molecular level for many years. Back in the 1930s, the American chemist Linus Pauling began to consider what kind of molecule an antibody must be. His proposals were somewhat off beam, but the idea of thinking in chemical terms about immunity was influential. Only in the 1960s were the general features of antibody structure revealed. We now know that antibodies are complex molecules, made up of both fixed and variable sections. That means they all have the same basic structure, but the fine details can vary, and it is this variation which makes possible the exactness of the lock-and-key fit with thousands of different antigens. Each time a particular antigen appears, exactly the right antibody is produced to meet it.

The details of how the right cells are selected within the body to make their antibodies at this moment were worked out in the late 1950s and earned Nobel prizes for the scientists concerned. But a fundamental theoretical problem remained. How was it that higher organisms were able to produce such an enormous number of antibodies? That question is still unresolved. Immunology is an area where molecular genetics has been particularly illuminating. After all, antibodies are large proteins, and they are specified genetically. Some molecular biologists have spent much time in recent years considering how the genes in antibody-producing cells are switched on and off to produce the right molecule at the right time. It is this kind of question that was being pursued in developing the monoclonal antibody technique.

So-called monoclonal antibodies are antibodies derived from specially created cell-lines, growing in culture. They produce just one kind of antibody and have been designed to create the desired antibody and no other. This was first done by Dr. Cesar Milstein and Dr. Georges Kohler at the MRC Laboratory of Molecular Biology in Cambridge, England, in 1975, in what was a digression from their main line of research.

Basically, monoclonal antibodies are created in hybrid cells, called hybridomas. Two kinds of cells are fused, because antibody-producing cells cannot survive in artificial cultures. They need to be cultured in this way in order to produce particular, pure antibodies to any one of a whole range of substances of interest. In a sense, the immune response is spread out to reveal its different aspects, so its incredible sensitivity, normally obscured by the diversity of antibodies produced by an injection, is put to use. In effect specific hybridomas are an analytical tool of immense power, since one can begin with a substance of unknown composition, make monoclonal antibodies to each of its components and then use each antibody to probe and analyse the substance of interest. Or one can simply take a known material and make hitherto unobtainable quantities of antibodies to it. This work now has exciting and wide-ranging implications. For example it has been used to treat leukaemia patients by making antibodies to particular antigens on leukaemic blood cells, enabling them to be removed from the blood.

It also has important potential in diagnostic medicine and research, in that monoclonal antibodies can be used as highly sensitive, very specific probes. For example, tumours are known to produce proteins that are characteristic of that form of cancer. The problem often is to identify the protein. If these molecules could be detected at a very early stage in the formation of a tumour (before there were sufficient cells to be seen on an X-ray picture) then treatment could begin much earlier, with a better chance of success.

Another example of their use, in a less extreme situation, is that of discrimination of different kinds of meat in a sample of mince. It is already possible to identify horse or kangaroo meat being passed off as minced beef in this way. It may soon be possible to do this with cooked hamburger meat as well. Or one can make antibodies to the proteins on the surface of blood cells and use them in identifying specific blood groups.

It has also been suggested that they could be used in developing new contraceptives, by making antibodies to proteins in human sperm. It is even possible that they might be sex-specific—that is, acting only on sperm carrying X chromosomes (female sperm) or Y chromosomes (male sperm). This would allow sex predetermination to be practised effectively at last, a prospect that has been regarded more and more critically, particularly by feminists, as its realisation approaches.

The list of already-realised applications of monoclonal antibodies is immense, and their commercial value in diagnosis, assays, purification and therapy is likely to be enormous. One estimate puts the diagnostic market at $870 million by 1985, ten years after the original paper.

When Milstein first achieved success with his experiments in 1975, these possibilities were but dimly perceived. Nevertheless, he wrote to his sponsor, the Medical Research Council, informing them that there could be some industrial applications, in the hope that the National Research and Development Corporation (NRDC), a government-run brokerage for inventions made in public laboratories, would help in the patenting and commercialisation of this work. It is worth saying that in his public statements Milstein has shown an endearing indifference to the idea that he might have gained personally from the commercialisation of his work. As a Jewish refugee from Argentina he gives every impression of being very happy to work in a distinguished academic laboratory and very absorbed in the activity of research as such.

In the event, for reasons that have never been made clear, the

NRDC did not act, and the key patents to work on monoclonal antibodies were taken out by American researchers. This created a great deal of resentment, and in the public discussions the finger has been pointed at the NRDC as the body concerned that failed to get things going quickly enough.

The whole business has become a *cause célèbre* and is represented as yet another failure to get British science, second to none in its incredible, breakneck inventiveness, turned into competitive products that will sell in international markets. Free marketeers conclude from this example that corporations must be allowed the most uninhibited access to university or government labs with the freedom to exploit anything that takes their fancy. Private enterprise is thought not to drop the pass like this. Reform-minded advocates of state intervention to promote economic growth tend to see the monoclonal antibody episode as a characteristic fumble by undertrained, overprotected bureaucrats. I am sure in this case we do not have the full story. One conclusion that I draw is that a particular style of research is now under threat, historical circumstances now having caught up with it. Research as the self-absorbed activity of a cloistered élite, used to having the polluting world of venture capital, patent litigation, advertising, marketing and portfolio analysis held at bay for them, is on its way out. As we have seen the membrane between these two worlds has recently become more permeable, and new influences are diffusing into the research laboratory.

One attempt to solve this problem has some of its roots in the monoclonal antibody saga. In 1980 the British Technology Group, which had been formed from the National Research and Development Corporation and the National Enterprise Board (NEB), set up a company called Celltech, its answer to the high-tech genetic engineering companies in America. The supporting capital came half from the NEB and half from private institutional investors. It has an exclusive right of access to research in MRC labs, like the one in which Milstein works. It is then supposed to be a major institution in the UK for commercialising research in biotechnology, performed with government funds. It is a kind of compromise between the free market buccaneering in America and the bureaucracy of the NRDC. Among its first products were several that exploited monoclonal antibody technology, including the antibody to interferon and a set of antibodies that allow blood cells to be typed by blood group more efficiently. (Cesar Milstein is one of the distinguished scientists on the scientific advisory committee of Celltech.)

In the spring of 1982 the Parliamentary Select Committee for the Department of Education and Science held a series of hearings on biotechnology. Amongst other things they reviewed the exclusive relationship that Celltech has with the MRC and concluded that it ought to be scrutinised very critically. The Tories on the committee were far from keen on this monopoly, and Celltech has now been told it will lose its exclusive rights over MRC research.

Failures on the scale of the monoclonal antibody case worsen the plight of disinterested research. Pressure for publicly funded research to be made into a facility for private industry grows stronger and more difficult to rebut. Equally, these pressures are changing research practices in academic institutions. The result is that biotechnology is being hailed as a necessary path that our future must tread and, simultaneously, is being removed from a visible public domain.

New Vaccines for Old Diseases: Sharpening the Immune Response

Viral Fractions

An immortal cell-line made by hybridising the tissues of a sick mouse, as Milstein did, is one way into the maze of immunity. There are others, which are refinements of established practices. In the 1950s molecular biologists like Francis Crick were in the habit of telling virologists, who were not eager to be advised by someone they perceived as an unskilled young upstart, to think in molecular terms. Think of viruses as structures built from a very economical set of instructions, Crick argued.

Since there is not very much DNA packaged inside the coat proteins of a virus, the genes cannot specify many proteins or very large ones. There just isn't room to encode the information, so the coat must be built from regular subunits, identical in shape, that stack together to make a shell. That is why viruses tend to look like spacecraft. Simplicity means geometrical regularity. It is this kind of thing a crystallographer would notice. We now know that the principle of economy goes even further, and that viruses have overlapping genes. In some viruses the same piece of DNA codes for more than one protein. It is like writing a paragraph and then dividing up the letters differently so as to produce another, entirely coherent, paragraph with a new meaning.

Now that far more is known about how viruses are put together some people's attention has turned to remaking viruses, by cutting out sections of their DNA (or their RNA, for those viruses that use RNA as their hereditary material). The objective is to produce a virus that lacks the genes that are central to some of its infective activities, but which still has the same basic structure and thus provokes the same antibodies as the natural virus. This would be immensely useful in vaccination. It turns out that the proteins on the coats of some viruses, or other fragments of the whole structure, will stimulate an immune response. The problem in making vaccines is often to ensure that the virus is inactivated or noninfective. Parts of viruses cannot be infectious, so they would be safer. The trick then is to get the part to stand for the whole. Interestingly, when this idea was first shown to work in 1976, the scientists concerned did not think to apply for patent protection.

Patent Struggles: Bad Faith and the End of Openness

By 1980, the situation had changed dramatically. The technique of constructing synthetic antigens formed the basis of a dispute between two research groups over priority of discovery and unacknowledged intellectual debts that may have a commercial value of millions of dollars. One group is led by molecular biologist Richard Lerner, working at the Scripps Clinic and Research Foundation. The other, led by Russell Doolittle, is based on the University of California at San Diego (UCSD) and at the Salk Institute. All of these centres are in or near La Jolla, a suburb of San Diego in California, so researchers are near enough to drop in on one another if they want some advice or assistance. From one such conversation a major row has surfaced.

Doolittle claims that he gave the other group an idea, and this group subsequently applied for patents on it. These in turn were used to attract investment from a pharmaceutical company, Johnson and Johnson. Not surprisingly, the leader of the Scripps group, Richard Lerner, strenuously denied all these accusations, and claimed instead that the novel use of the synthetic antigen technique, about which both groups have published papers almost simultaneously, came to him independently. He maintains that he thought of the idea ten months earlier on a walk with his colleagues in Central Park, New York, while attending a conference.

He also says that he didn't mention all this in the critical conversation with Russell Doolittle, because he didn't want to embarrass him by making it clear that work of which Doolittle was obviously proud had been thought of before. And he claims that he has documents that prove that his lab was already moving on this before he came to Doolittle for a friendly chat.

For his part Doolittle suggests that there is an element of ambiguity and evasiveness in some of the statements that Lerner has made and that the documents (an order for a research peptide from a small chemical company) are not conclusive. Furthermore Doolittle claims that the helpful role that others played in the work of Lerner's group is grossly underplayed in their publications, something that would not be irrelevant to the success of a patent application.

Crucially both groups say that they saw the commercial use of the idea of making bits of virus as a way to make safer vaccines before publishing their descriptions of its use as a research tool. The Scripps group does have a patent application in train and is involved in a commercial venture to manufacture synthetic vaccines. The University of California will not reveal whether they are also applying for a patent.

It is an amazing saga, reminiscent of a murder story, with some documents available for inspection, some withheld in commercial secrecy. Both sides have a "story" that accounts for their and the other's actions. Both have modified their descriptions of how they acted as the affair has proceeded. Doolittle apparently was prepared to overlook what he saw as scientific bad manners, the failure to acknowledge freely given research material and assistance, until the announcement of the link with Johnson and Johnson was made.

It is clear that Lerner and his group learnt from the conversation with Doolittle that they were in a race to publish an idea, and that by saying nothing Lerner left his competitors without this information. It is a graphic example of what commercial pressures can do to exacerbate problems of openness and generosity. The effect is likely to be to act as a powerful inhibition on communication and exchange of materials in research. So much, one fears, for "the scientific community" and the "disinterested pursuit of truth" which form part of the appeal of scientific endeavour.

It is now quite usual and apparently acceptable for researchers with commercial connections to sign agreements that bind them in the use and distribution of research materials and specify what kinds of collaboration are permitted. Genentech has its patent lawyers attend research seminars in the company, so that the patent claims can be shaped around the research as it emerges. To utter an idea at

Genentech is therefore to have its patentability assessed. Journal editors are now being forced to record with meticulous care the exact time and date when scientific papers arrive at their offices because US patent law allows one year after public disclosure within which a patent application may be filed. In Britain this year's grace does not exist, so clearly anything revealed there must be prescreened for its commercial value. These are major changes in research ethos and practice.

A World of New Vaccines

The technique of producing synthetic antigens lies in the heart of various programmes to produce new vaccines for serious and hitherto intractable human diseases, like the jaundice caused by hepatitis B and the venereal conditions and cold sores caused by the herpes viruses. There is also great interest in animal vaccines. The market for a foot-and-mouth disease vaccine is put at $300–500 million in the US, and it has been suggested disparagingly that animal vaccines are the preferred target, whilst other less profitable conditions affecting human beings are ignored. It is technically possible that better vaccines may be produced for diseases for which vaccination is already possible, like polio and rabies.

One of the most striking examples of a disease for which vaccination could soon be an answer is malaria, a condition that has provoked a variety of different counter-strategies over the last hundred years, including antimalarial drugs, DDT spraying, irrigation, insecticides and the breeding of sterile mosquitoes. But one problem is that an enormous number of people would have to be vaccinated, although the example of smallpox suggests that it could be done.

On the other hand, the vaccines have to be stable, since it takes time to get into remote regions away from the metropolitan centres of distribution. In tropical countries the heat has often broken down the vaccine by the time it is injected so that, unbeknown to those concerned, it is ineffective. No one knows how stable the new vaccines will be.

Without appropriate refrigeration technology, the same problems are likely to remain, as will the lack of sanitation, adequate food, sufficient fuel and housing materials that exacerbate the disease problems of something like a third of the world population. This is fitting background against which to consider a final development of applied genetics, which the technical ability to move genes around into specific organisms brings closer, namely the treatment by gene transfer of human genetic disease.

Gene Therapy: The Itch to Be First

Recombinant DNA research, the technical revolution in molecular genetics of the early 1970s, spawned many of the experimental procedures that allow one to conceive and design a vast range of new industrial processes. It has also led to a tremendous increase in knowledge of the human genetic system. The detail of that knowledge and the technical skills that created it have allowed scientists to push onwards to do human genetic engineering or gene therapy, as it now tends to be called. Not only are scientists almost in a position to make that step—indeed, at least one has already tried—but it is very obvious that powerful forces are accelerating them down that road.

What seemed five years ago a rather distant, slightly disreputable, controversial and embarrassing topic appears now to many scientists in a different light. Now that it is so near, gene therapy is becoming technically interesting, and a valid object of study for scientists looking for professional recognition. It is almost fashionable, and the itch to be the first person to manage it successfully is getting stronger. The point is that professional attitudes have evolved in a surprising way, which it is important to recognise. What is striking is the shift of tone from alarm, embarrassment and strong aversion of only five years ago, to the present real interest in the technicalities of a new medical procedure. It is not disreputable for a scientist to dabble in gene therapy. As it happens, I don't feel that some awful barrier is about to be crossed, that we are on the banks of a moral Rubicon that will lead to the fall of a spiritual Rome. But I do feel that gene therapy is very sophisticated resource-consuming, experimental medicine, in which the clinicians' interests are involved as much as are those of the patients, and I want again to raise the question of whether this should be on our list of priorities now.

Genetic diseases tend to be rare, but significant in aggregate, as you add up the numbers of people involved. The figure given for the combined incidence in a country like Britain is around two per cent of all live births or roughly 20,000 babies a year. Of these, about half are major chromosomal abnormalities, like Down's syndrome, and hereditary structural or anatomical abnormalities, such as cleft palate or hole in the heart. Some of these can be corrected by surgery. In conditions like Down's syndrome, we know that it comes about because of the acquisition of an extra chromosome, an extra No. 21 in a set of twenty-three pairs in normal human beings, somewhere along the road from the formation of germ cells to fertilisation, but we do not know in any more detail how the condition is caused.

With others, the single gene defects, which together make up about one per cent of live births, we know far more about the links between the presence of a gene in a particular form and the resulting medical problems. In sickle-cell anaemia, for example, we know that a single alteration in the gene for haemoglobin is sufficient to reorder the amino acid sequence of two of the globin chains. This is sufficient to cause the haemoglobin to form long thin crystals when the amount of oxygen in the bloodstream is limited, so that the red blood cells collapse into a crescent-shaped configuration. The result is that the tissues fail to get the oxygen they need, blood vessels get blocked and damaged and the person concerned experiences a great deal of pain.

Sickle-cell anaemia is very specific in its cause. There is a related condition called thalassaemia, which arises from any one of a whole cluster of different structural defects in the globin genes. It has been called the world's commonest genetic disease, affecting some millions of people around the world. One estimate is that 200,000 children die each year from sickle-cell anaemia and thalassaemia. Although we now know a great deal about how particular kinds of thalassaemia are caused, we do not know how to treat it, and people who suffer from it are likely to die in infancy or adolescence. Genetic diseases quite often have that kind of severity. They tend to be physiological disorders of some sort that rapidly lead to the catastrophic failure of particular organs, tissues or systems.

One possibility for some genetic diseases is to practise a form of prevention that takes advantage of the fact that they are genetic. Human chromosomes come in pairs, and at corresponding positions on each chromosome of a pair, the gene concerned with a particular trait or process may take a form slightly different from that of its partner on the other chromosome. Thus in the case of sickle-cell anaemia you need to acquire two abnormal globin genes, at the same site on each chromosome, for the problem to arise. With just one copy of the sickle-cell gene, quite enough normal haemoglobin gets made.

Conditions in which two copies of the gene must be inherited for problems to arise, we call "recessive." It is possible to identify normal healthy individuals who carry a recessive gene for, say, thalassaemia. The significance of this is that although they are unaffected, if they have children with someone who also has a single copy of the same thalassaemia gene, then there is a twenty-five per cent chance that those children will have thalassaemia.

We can ignore the reasons why the chance is twenty-five per cent. The point is that with that kind of foreknowledge of serious problems, a number of choices open up if they wish to have children. One is to take the risk anyway, regardless of the outcome. Another is to conceive and then have prenatal diagnosis performed, which can disclose whether the foetus is affected by thalassaemia, so that the pregnancy may be terminated. In fact, in the case of thalassaemia this diagnosis is not particularly easy or safe to do, but it is becoming technically easier. It is also likely to be psychologically demanding, and for some people it is difficult or unacceptable to justify morally. Another option is to find another form of procreation that evades the problem, such as artificial insemination by a donor known not to carry the relevant gene. Or one can adopt children, or decide not to have them, or find another partner. None of these decisions are easy to take, but many people find them preferable to having one or more of one's children die of an untreatable disease.

The more research is done in human molecular genetics, sequencing particular genes, identifying where they are located on which chromosomes, linking them to particular gene products and metabolic pathways, the more information is available with which to identify unaffected carriers of particular conditions and on which to base genetic screening programmes. More and more conditions will be diagnosable in this way, both in adults and through prenatal tests. That has to be a good thing, even though screening programmes have created a number of problems. One of these is the thoughtless creation of anxiety and the stigmatisation of perfectly healthy people as somehow defective. Medical enthusiasts have to learn that the meaning of a particular condition for them may be quite different from the significance for someone of a different age, race, marital status, social class and educational background.

I have represented the identification of carriers of a recessive gene for a condition as a matter of individual choice, something people can opt to do, if they have reason to think that the disease may run in their family. Another possibility is to offer this medical service to whole groups of people, and this is called genetic screening. Usually it is made available on a voluntary basis to groups at risk, but in some cases the need to have people's genetic status ascertained has been made compulsory. Some American states have laws that require the screening of black schoolchildren for sickle-cell anaemia as a condition of attending high school. Whilst the intention in making screening mandatory is to ensure that people get essential genetic information, the potential for racist abuse is considerable.

Those setting up screening programmes tend to be white, married, highly trained, sexually experienced males in secure, high-status employment. Those whom, in all sincerity, they are trying to help, tend to be unmarried, or without sexual partners, from ethnic minorities, from a lower social class and to have had far less education.

A consultant haematologist recently conducted a small screening programme among the children of various ethnic groups in an inner-city school in Britain. One girl was seriously distressed by what she was told. It shouldn't have taken much imagination to see that all kinds of problems, of stigmatisation and anxiety, those that can be exacerbated by the kind of social pressures that work their way through schools, needed careful consideration that was missing. This kind of thing is likely to happen again and again in the 1980s.

Finally, we come to human genetic engineering or gene therapy as it is coming to be called. This is an attempt at cure rather than prevention. Basically it means attempting to correct a medical condition such as thalassaemia by introducing into specific tissues DNA molecules containing a normal copy of the gene whose malfunction is causing the problems. The idea is that if the gene can be inserted into the relevant cells so that it operates correctly, then the normal bodily functioning, missing since the onset of the disease, will be resumed. The technical term for this is somatic cell modification. To do this properly is incredibly difficult. One man has tried it already and failed. Twenty or thirty research groups around the world are gathering themselves for more attempts at some point in the near future. The technical capacity does not exist to attempt gene therapy yet, but it cannot be far off. An alternative is to insert the gene in multiple copies into the fertilised egg or the early embryo. The successful attempts to get rabbit globin genes taken up into mouse cells mentioned in the first chapter fell into this category. This is called germ cell modification.

From one point of view thalassaemia is an ideal condition with which to attempt gene therapy. It is a serious single-gene defect that is untreatable. The primitive red cells of the bone marrow are accessible by the painful procedure of taking a sample and can be retransplanted into the bone. The gene system is well understood, its chromosomal position known, and relevant DNA sequences can be made from normal human cells. On the other hand, not enough is known about how to get the genes into marrow cells in a safe, controlled and effective way.

By 1980 a haematologist from Los Angeles, Martin Cline, felt that his experiments with mice gave him sufficient knowledge to attempt to treat thalassaemia patients by gene therapy. The application to make this step was eventually turned down by local and national committees whose permission was necessary. Instead he made arrangements to do the same clinical experiments in Israel and Italy. A BBC television programme on this subject made clear that in setting up the treatment for the Israeli patient he misrepresented his experimental technique to the hospital authorities. When the details of his work were made public, he was forthrightly condemned by his colleagues and asked to resign one of his professional posts. What disturbed people was that he should have felt able to extrapolate with confidence from mice to human beings, when the overwhelming weight of his peers' opinion was that his procedures had no possibility of success.

It is not for me to speculate in print as to his motives. I can only say that in his television appearance he simply stressed that in working with chronically ill patients he felt impelled to take risks that might appear unacceptable to observers less involved in clinical research. That may be so, but it does not justify ignoring the informed deliberations of ethical review committees. Equally, it is hard to see how one can justify this and other kinds of highly specialised, expensive, superstar experiments when many thousands of sufferers from the blood disorders have much more prosaic medical needs that go unfulfilled in their short lives. Is this unseemly haste going to become less unusual as career and commercial pressures grow in biological and medical research? I think so.

Human cloning will fit the same pattern, with an innovative practitioner ignoring scores of more mundane medical problems and slipping past the controls of his or her peers to help a few people with just enough cash to buy that kind of service. One reason for not saying more about cloning in this book is my belief that the moral outrage that surrounds such work is a distraction from the structural and economic dimensions of health and disease. It is not that these violations of basic taboos are unimportant, but that the systematic channelling of resources into profitable areas that benefit only a minority is far more scandalous.

In Conclusion

We've covered some ground since the beginning of this chapter. Making clotting factor in bacteria is a different exercise from clinical gene therapy. Both, as it happens, are approaches to the treatment of haemophilia. The former is being done; the latter lies in the future and may never be used if clotting factor can be made cheaply and haemophilia be diagnosed antenatally more precisely than at present. Both belong to the family of technologies that we label "biotechnology," as do all the other things in this chapter: bacterial insulin, hybridomas, synthetic antigens, contraceptive vaccine and immune assays for blood typing. Each of them has arisen from a specific set of medical, technical and financial needs. Each of them embodies choices over health priorities which are open to question, to say the least.

Medical biotechnology is the beginning of the agenda. It is here that the revolution will begin. Already there are products on the market produced by new techniques and there are plenty of ideas for more. The excitement at the range of possible medical applications is intense, and much effort is being expended trying to get established on the new terrain. For many entrepreneurs the ultimate dream is to become a big drug company. One or two concerns have made startling progress already in that direction.

There is something quite impressive about the rate of growth of medical biotechnology companies. No doubt behind the scenes there is far more anxiety, aggression and incompetence than the public image is allowed to suggest. But the picture of research projects being created and combined at a furious pace, of things being made to happen, has a certain appeal. Similarly in larger companies with a more measured tread one cannot ignore the pride from some of the team leaders in the pace of achievement. But at the end of the day, whom will all this activity help? Some people certainly, but, I suspect, they will only be that minority already well supplied with medical goods and services, and the costs of providing cures for the ills of this social stratum will not fall. Too much money is made from putting molecules into people at a profit for very much to change without major convulsions.

That, for me, tarnishes in the gleaming image of biotechnology. It is not that most people in health-care companies don't care that their work never touches the lives of millions of people. It is just that they have different priorities: an adequate return on capital invested. Their view of medicine is based on the view that people must first accumulate wealth and then buy health; otherwise there is no deal. The public good may take too long to be a worthwhile business proposition. "Pure" research is used to fill that gap, but it is being increasingly subject to commercial criteria.

5

Speed-up at the Plant

Agriculture is something like ten thousand years old. It was the first major change in the way that organised groups of human beings acted on their environment to produce food. As soon as it began, farmers started to select and improve their crop plants and their livestock, to increase the amount of surplus produce from their land. Plant and animal breeding are at the heart of agriculture. Their histories interlock. That constant activity of deliberate selection, controlled breeding and improvement is still central to agriculture throughout the world, whether in the developed or developing countries. Now biotechnology is giving this process a new twist.

This chapter is devoted mainly to plants; the animals are penned in a section of their own in the middle of the chapter. Plant breeding is not just technical improvement. It is crucial to the constant restructuring of agriculture, which ties an ever smaller number of producers to the food processors on the one hand, and the fertiliser, equipment and fuel companies on the other. The task of making new plants is fundamental to the continued remaking of agriculture. The latest developments in plant biotechnology allow the industrialisation and concentration of agriculture to be driven forward even more quickly. It will accelerate control over agriculture by corporations that service it, invest in it and consume its products. In their hands lies the future of the world's food supply since they control the resources like fertilisers and pesticides that are now essential to maintain the yields. They are coming to control the living resources of plant genes, from which new varieties will then be bred.

Modern agriculture has produced an abundance of food for the fortunate few. Major crop-producing regions of the developed world consistently produce enormous surpluses of cereals and dairy products. The butter mountain of the EEC and the sales of American, Canadian and Argentinian grain to the Soviet Union are two examples. There are, of course, powerful economic forces that encourage such levels of production, such as guaranteed prices for farm products.

Without the technical means to get so much from agricultural land, this level of "overproduction" could not occur. The yields of most crops in the developed world have been increased dramatically in the past fifty years. For example, in the Corn Belt of the Midwest you can harvest more than a hundred bushels of corn from an acre of land where twenty bushesls would have grown in the 1930s. Some of that increase has come from the introduction of higher-yielding varieties, but realising their greater potential has meant the use of enormous amounts of artificial fertiliser. To make that takes energy, and energy costs money. It has been calculated that energy inputs for corn production in the USA increased by over two hundred per cent between 1945 and 1970, with fertiliser accounting for over a third of the total. The production of fertilisers in the United States takes up three per cent of natural gas demand, and the energy cost of distributing the fertiliser is about the same.

While energy costs continue to rise, the land available for agriculture gets smaller. One million or more acres of US farmland each year disappear beneath houses, roads, factories and shopping centres. Moreover, an increasing proportion of some crops such as corn, sugar cane and cassava is being turned over to produce energy in the form of alcohol and is not being used for food. To maintain current levels of food production, more has to come from less land, while holding energy inputs down. That won't be easy to do. The fact that we, in the developed world, don't experience this as a crisis is because of the massive subsidies to agriculture, the abundance of food and the strength of agricultural interests.

These trends mean that big farmers in the industrialized countries will be squeezed, although an enormous amount of research will be done to help them cope. A tremendous amount of money is made out of the sale of seeds, pesticides, fuel, fertilisers and farm equipment, and the corporations concerned don't want to see that business go down the drain.

The contradictions between the need to produce more food and the costs of doing so can be condensed into a problem for the plant end of the food production process. That is where scientists are being asked to speed things up. Faster ripening, faster seed filling, synchronised growth to facilitate mechanical harvesting, more efficient use of sunlight—all these are goals for plant breeding. If only ears of corn would form more quickly, if only photosynthesis in wheat went as rapidly as in sugar cane, then many of the problems would shrink. Similarly, if breeding for greater tolerance to pests, or an ability to withstand drought or flooding were more successful, some of the pressure could be relieved. If plants could be created which were more tolerant to salt, then land that has been overused (or that lies over salty underground springs, or that can only be irrigated with sea water) could be opened up. One estimate puts the global acreage of largely unused saline soils at 3.8 million square miles, compared with the 6 million square miles that make up present croplands. The possible increase of useable land is nearly two thirds.

To develop a new garden pea or a better broad bean takes time and labour, and here industry is trying to accelerate, pushing for new techniques that allow breeders to create new varieties, even new *kinds* of plants more quickly. The basic idea is to work with plant cells or tissues, to manipulate them to order, and then to allow them to grow into mature plants. A more radical possibility is to do away with plants as organized systems altogether. As I mentioned in chapter 1, plant cells can be grown in culture, to produce licorice, or nicotine, or codeine. But whether this method could be used for food plants is not clear. Certainly there are many plans to culture protein-rich bacteria and algae in all kinds of media, and for some people these will be a food of the future.

All this has been called, in anticipation, "the second green revolution." In one way, this is a quite useful description, because it refers us back to the implications of the first so-called green revolution. This was an attempt to "improve" the performance of Third World agriculture by a kind of technical fix. High-yielding varieties of crops, such as wheat and rice, produced by an international network of research institutes, were introduced as a means by which "progressive" development could take place. These new plants were conceived from within an industrialised agricultural system, however; and a dependence on the products essential to that system was introduced into the Third World along with the new varieties. Not only was this "revolution" a relative failure, but the induced dependence has had profound social and economic effects, as this chapter will later show. At its simplest, it required the import of massive and expensive amounts of fertiliser and equipment. Part of the aim of the "green revolution" was to squeeze out the "inefficient" small farmer.

If this was the effect of the first revolution, then what the second may bring will take the Third World even further down the road of dependence. This is not merely an unfortunate by-product. It is a measured step in the expansion of an industry to global scale. For about seventy per cent of the world's farmers, who cultivate less than one hectare, these present developments spell out an even harder struggle to survive. They are, in truth, at the sharp end of this transformation of agriculture—a transformation being shaped by a more radical alteration of the plant itself.

So, rather than call this a second green revolution, I would prefer to think of what is happening now as an attempted speed-up at the plant. Plant biotechnology is really a series of manoeuvres by corporations that sell or use plants, designed to get more plant for a given input of money. As we will see, many of the techniques resemble those being used in the medical sector—cell fusion, gene transfers, and selecting for particular cell types—although many of the puzzles seem harder to solve. Plants are more complex biological systems. Many more genes control single functions than in bacteria.

The pattern of industrial interest is also similar. There are small research companies setting up, with spliced-together names such as Calgene and Agrigenetics and Zoecon, and there is active interest by big corporations such as Atlantic Richfield, Monsanto, Du Pont, Stauffer Chemical, Occidental Petroleum, Exxon and ICI. As we shall see, the problems of that kind of involvement are similar too. The corporations go for the projects that are profitable for them, and that reproduce the same kinds of agriculture. To do that they are claiming new plants as their property. The patenting and investment in university research is restricting the openness of academic results and enquiry.

Here once again we are dealing with power. In this case the power in question is the power to exploit the genetic resources of plants in order to gain control over future markets. The developing ability to design, create, and patent specific kinds of plants will confer upon the suppliers of plant varieties a greater degree of control over what is grown, over what substances are bought to protect or increase yields, over the price at which seeds are sold and over the purpose for which crops are grown. Through the design of new plants a new structure of dependence on agri-business firms is being planned, in return for which some of us will get food. The savage irony of this is that many of the genes used to create the new plants will have come from those very countries where food is short. The genetic resources of such countries, which form a large part of the world's ever-decreasing number of species, will be used abroad and patented, or offered back to them in new plants, at prices they can ill afford.

I propose to examine six areas in which biotechnology is significantly affecting agriculture. These are, first, the production of new kinds of plant; then, the consequent pressure on nitrogen, on which plant growth depends; then, the standardisation of trees. A fourth section examines work on animal productivity; a fifth, the squeeze on the productivity of human farm workers, through mechanisation. The last, in many ways the most crucial, section analyses the battle for control over seed. Throughout, we shall glimpse the beginnings of an epochal struggle over food and ask what kind of resource will food be? For whom?

New Plants for Old

Some of the new plants are spectacular in name and concept, but pretty ordinary to look at. There are the hybrids of plant and animal cells, which represent an extraordinary fusion of species, but which look like ordinary plants. The difference lies in their nutritional value. The addition of animal cells means that, hopefully, mammalian proteins will be produced by the plant. This will boost its worth as a source of food. For instance there is the "pomato," a cross between the potato and the tomato. At the moment, it produces leaves and roots, but apparently little else. One day it might have ripening tomatoes on top, and potatoes below. To the peril of any Colorado beetle in east Anglia, other work is going on to cross potatoes with plants that digest insects. The idea is that crop potatoes should develop the leaf hairs that secrete powerful chemicals which can turn insects into liquid blobs.

How are these remarkable feats carried out? An important precondition is the ability to break down plant material into a primitive cellular form by dissolving away components of the cell wall with enzymes, leaving cells that still work but are in a rather naked state called protoplasts. In this condition the cells can be subjected to various kinds of manipulation and be fused with protoplasts from other species. This is not unlike the creation of hybrid mammalian cells, such as the hybridomas considered in the last chapter that produce monoclonal antibodies. Protoplasts can be persuaded to regrow a proper cell wall and to start dividing so as to form clumps of cells. If the clumps are suitably nurtured and soaked in hormones they reorganise themselves to produce shoots and roots and become tiny plantlets that will gradually grow into adult plants.

One example of this technique is the work on disease-resistant potatoes by James F. Shepard and his associates at Kansas State University. The team spent a long time working out how to get potato tissue to grow in culture, and which hormones and growth substances would cause the protoplasts to form clusters and develop into whole plants. While doing this they discovered a surprising degree of variation amongst potato protoplasts. Some of them were more disease-resistant than others; some grew more rapidly or produced more uniform potatoes. The immediate implication was that the team had a method of selecting for particular traits in the laboratory. They exposed the growing cells to toxins—poisons—from the fungus that causes potato blight, and selected out those that stood up best to this form of test-tube persecution. When the resulting plants and their descendants were tested, it was found that they were resistant to potato blight. Discounting the years that were spent in developing a way of getting whole plants from protoplasts, Shepard's group compressed into a few months work that would have taken several years of selection in the field. Shepard has gone on to work on the sweet potato. He has also developed a commercial relationship with a new plant genetic engineering company, called Advanced Genetics Science Ltd.

What can be done with potatoes can also be done with carrots and tomatoes. Indeed the first clonal carrots, created by the propagation of cells from a single plant, were produced in the early 1960s by F. C. Steward at Cornell University. Growing cereals such as wheat and rye from protoplasts has proved more difficult. Nonetheless when you visit a plant genetic engineering company they are likely to show you with some pride, their "wheat field of the future," rows and rows of tiny plants growing in culture flasks in a small room under artificial light. It is something on which they are working very intensively indeed. It is not hard to understand why this might be so, if we consider the relative production of different crops around the world. Cereals dominate. The market for cereals is the biggest and most obvious target for plant breeders.

The Nitrogen Fix

This section isn't the easiest to read, but the issues it raises are very important indeed. Biotechnology might be able to make plants relatively self-sufficient for fertilisers. The world's deserts might bloom. To explain this possibility, it is necessary to talk about the main fertilising element—nitrogen.

Agriculture, at present, is locked into an industrial system that stretches from the breeding and production of high-yield plants and seed, through the fertilisers that are needed to feed them and pesticides to protect them, to the processing of food. Biotechnology is now being used at the plant end of this system to shackle agriculture in this posture of dependence. For large corporations, mostly from the chemical industry, this represents a way of diversifying out of a crowded sector of production that is encountering its own crisis. It is a strategic move designed to create new markets for the output of the chemical industry. In chapter 6 we consider new ways of creating inputs to the chemical industry, and this brings us back to plants.

In this new state of things, the plant becomes the locus of an industrial system. This involves no structural changes in energy-intensive, mechanical agriculture. Rather, biotechnology is being used to implant this system ever more deeply. By moving new genes into plants, an industrial sector is fixed in their very cells. A prime example of this is the attempt to exploit the nitrogen-processing capabilities of certain bacteria.

Increasing crop yields has meant an ever-mounting need for fertilisers. Nitrogen is the key resource needed to satisfy the demand that these plants make. Its significance as a raw material can be measured by the fact that every amino acid, the basic element from which proteins are built, contains at least one nitrogen atom. Any living organism, therefore, needs nitrogen to grow. It is an abundant element of the earth's atmosphere, making up about four fifths of the air we breathe. Yet nitrogen is inert and defiantly inaccessible as a chemical. The two constituent atoms of molecular nitrogen are linked together by a powerful bond that takes a great deal of energy to break. So although every organism above ground is continually immersed in nitrogen, it can only be assimilated by an indirect route as "fixed" nitrogen. This is nitrogen that is combined either with oxygen (nitrates) or with hydrogen (ammonia).

The movement of nitrogen between soil, atmosphere and plants is cyclic with a constant ebb and flow of assimilation and release. Organisms grow by assimilating fixed nitrogen from a reservoir in the soil. This is released back again through their excretion and decay. In the soil itself there are constant flows into and out of the reservoir. Two kinds of bacteria aid the flow in. Organic remains are broken down with the release of ammonia, which is transformed into soil nitrates by nitrifying bacteria. Other nitrogen-fixing bacteria, either living an independent existence or bound in symbiosis with a particular species of plant, take up soil nitrogen and convert it to a form that plants can assimilate. Those living in symbiosis with plants trade their nitrogen for the plant's energy, to their mutual advantage. From the air some fixed nitrogen is returned to the soil through lightning discharges, which enforce the union of nitrogen and oxygen. Denitrifying bacteria aid the flow outwards by breaking down nitrates to release free nitrogen, which escapes to the air. Rain flowing through the soil takes nitrates into the rivers.

The growth of high-yield crops can quickly deplete the soil reservoir. Hence the need for intensive fertilisation and the mass production of artificial fertilisers. In the nineteenth century, the bulk of artificial fertiliser was made from vast deposits of guano, the mineralised form of seabirds' excrement, rich in ammonia. Its principal source was Chile. Access to remote places where guano was found was aided through an extensive territorial and trading empire. As tensions between the imperialist powers increased in the late nineteenth century, those such as Germany without easy access to raw materials in their colonies, and with an innovative chemical industry, began to look for new ways of making fertilisers. Just before the First World War, two German scientists, Fritz Haber and Karl Bosch, developed a process for making ammonia by combining nitrogen and hydrogen at high temperatures and pressures in the presence of a catalyst. That process still underlies the production of artificial fertilisers, although the requirement for hydrogen, derived from natural gas or oil, links the costs of production to the costs of fossil fuels.

Nonetheless, nitrogen fertiliser continues to form the bulk of all fertiliser used, and the tonnage is still increasing. Between 1950 and 1980 the total amount used in the US has gone up by twelve times. Moreover in 1970 over three quarters of the world's use of fertiliser occurred in the US and Europe. Behind this imbalance stands the shadowed history of the Third World. Dependence on the purchase of industrially produced fertilisers has taken the place of an imperial domination that once turned productive land in the colonies over to cash crops such as cotton, coffee, sugar and rubber. New industrialised agriculture is being inserted into Third World economies at a price well beyond the means of the nation states concerned.

However, only about one quarter of the total world production of fixed nitrogen involves the Haber process. Virtually all the remainder, an estimated 150 million metric tonnes per year, is a bacterial product. Most of this is the unwilled legacy of nitrogen-fixing bacteria already in the soil. Some of this total, however, comes from the deliberate use of plants such as peas, beans, clover and alfalfa, grown to enrich soils depleted by the cultivation of cereals. Restoring nitrogen to the soil is the main reason for crop rotation. The usefulness of such plants has been known for centuries; but why, and in particular, what role nitrogen has in this has only more recently been discovered. Only in 1888 did the experiments of Hellriegel and Wilfarth show that nitrogen was fixed by the symbiotic association of these leguminous plants and bacteria. To gain a purchase on the planned manipulation of the processes of nitrogen fixing, it is worth looking at the way this happens in the case of the legumes. Legumes have proved accommodating to various species of *Rhizobium* bacteria. Their mutual arrangements for nitrogen fixing are among the most sophisticated.

The *Rhizobium* enters the legume root through a root hair (a cell on the surface of the root which is specialised for absorption). The root hair turns itself into a conduit for the *Rhizobia,* which move inwards in a thread towards the cortical cells of the root. If the microbial infection spreads that far, the root swells and a nodule forms, which becomes full of millions of nitrogen-fixing bacteria. This process of "infection," a nonfatal, mutually advantageous invasion of the legume root system, is highly specific. Particular species of *Rhizobium* are recognised and admitted by each legume, and only that bacterial species can colonise a particular plant.

The sequence of reactions by which soil nitrogen is converted to ammonia in nitrogen-fixing bacteria is very complex and still not completely understood. The basic enzyme involved is called nitrogenase, which catalyses the break-up of the strong double bond of the nitrogen molecule and its subsequent reduction to ammonia. This process of nitrogen fixation is energy-intensive. Some of the energy that the bacteria require to carry it out comes from their symbiotic partner, the plant within which they are living. In a bilateral arrangement between energy for nitrogen-fixation are traded nitrates essential for growth. Mutual advantage is the essence of symbiosis.

One option, in exploring the boundaries of nitrogen fixation with a view to extending them, is to see if any specific combinations of bacteria and plants are particularly outstanding. It turns out that some strains of *Rhizobia* are more effective than others, so screening for superfixing strains in the lab makes sense. (One problem is that these prodigies don't always do well in the real world of the soil, being outperformed by the established all-rounders. The answer then is to cross-breed.) This systematic survey of bacterial performance has shown that some nitrogen-fixing bacteria have a back-up biochemical system that utilises the spare hydrogen released by the main fixation process. These so-called hup [+] strains are more efficient, and their use can enhance crop yields. One possibility being examined is the transfer of the plasmid carrying the hup [+] genes into other bacteria. These ideas all seek to enhance the performance of plants that will already accept nitrogen-fixing bacteria. *Rhizobium* inoculants are already a commercial product that farmers growing soya beans can buy. To achieve their full potential they need to be matched to local soil conditions and farming practices. UNESCO is funding a general programme to train microbiologists in less developed countries and to develop *Rhizobia* that could be used in peasant agriculture. While on the face of it, this seems a progressive idea, it could also prepare the ground for a new range of agricultural products to be sold to the Third World.

Another possibility is to grow nitrogen-fixing bacteria, particularly the free-living ones, on a spare carbon source and then use the resulting bacterial sludge as green fertiliser. Something like this already occurs in rice paddies where the water fern *Azolla* is grown. Its leaf pores are colonized by an alga, *Anabaena azollae,* which fixes nitrogen. The plants are allowed to proliferate in the paddy and are then ploughed into the mud to nourish the roots of the growing rice. This had been done in Vietnam and China for hundreds of years. Intensive research is now going on to develop this technique. A more radical possibility for cereal agriculture is the idea of creating new varieties that will accept nitrogen-fixing bacteria as symbiotic partners.

The specificity of recognition between a particular legume and a particular micro-organism is genetically controlled. The plant makes identification molecules that say to bacteria, "I'm clover" or whatever, and bacteria from a given species, *Rhizobium trifoli* in the case of clover, know how to recognise it. But it is possible by genetic manipulation to get bacteria to recognise, colonize and form nodules on plants with which they do not normally associate. The plant makes the identification molecules from another species. The fact that nitrogen-fixing bacteria have been found that live on the roots of tropical grasses is interesting, because these grasses are distant relatives of wheat and rye. Possibly getting such bacteria to grow on the roots of new varieties of wheat and rye might not be too difficult. In this case bacterial inoculants, which would be cheaper to produce than artificial fertiliser, could substitute for it as a product. To keep up their yields farmers would have to buy the new varieties that accepted these root bacteria and the inoculants with which to spread their fields.

Finally there is the very ambitious project of producing nitrogen-fixing plants. Bacterial nitrogen fixation is controlled by a set of genes, the *nif* system, which specify the required enzymes and control the amount of each that is made. The transfer of the entire *nif* system on a bacterial plasmid into *E. coli* has already occurred, and *E. coli* was turned thereby into a nitrogen-fixing organism. Similar experiments with yeast, an organism with a different cellular machinery, have proved less successful. Despite this some people are still working at the transfer of sets of genes, like the *nif* genes, into plant cells, using plasmids or viruses as vectors. Moving the genes around is not that difficult; getting them to function is the real problem.

The basic technique is to use a plasmid from the micro-organism *Agrobacterium tumefaciens* which causes crown-shaped tumours in a wide range of plants. The genes responsible for this plant tumour are carried on a plasmid in the bacterium, called a Ti-plasmid. When the bacterium infects a plant, the plasmid DNA is incorporated into one of the plant chromosomes in the infected cells. These cells are turned into a plant tumour, which proliferates rapidly. It has proved possible to splice extraneous DNA into Ti-plasmids and get the combined genetic elements incorporated into plant chromosomes. Small slices of tissue can be taken from the resulting crown gall tumour, which can be made to regenerate into whole plants. The chromosomes of these plants then contain the foreign genes spliced into the plasmid DNA. Another vector in use now is the cauliflower mosaic virus, which will also move pieces of DNA around. It might be possible to use this technique as a way of speeding up the production of new varieties, for example with a more nutritious protein content or greater photosynthetic efficiency. The problem is that qualities like this are controlled by a considerable number of genes and transferring an entire cluster in a working condition is likely to be formidably difficult.

The obvious implication is that this research is unlikely to have any impact on agriculture for a considerable while, a decade at least and perhaps longer. The other less ambitious routes may have an effect rather more quickly. There is however a great deal of corporate interest in plant breeding at the moment, indicating that in the longer term the corporations concerned, principally from the chemical industry, see major opportunities in selling new plants. Their move into this area is partly a response to technical developments in plant science, partly a way of diversifying out of the crowded chemical sector and partly due to commercial opportunities presented by changes in patent laws. It is now easier to own a plant species than it used to be. If the plans that we have been discussing come to fruition and result in new plants that offer significant economic advantages, then they will be put on the market as the private property of agribusiness corportions.

From the point of view of those that sell fertilisers and pesticides, it is becoming increasingly important to control what kinds of plants are grown, because of a tightening correspondence between the packages of chemicals that are sold to farmers to increase their yields and the plants that respond to them. If in the future some of these chemical products are replaced by other products, such as bacteria, or even engineered into the crop plants, the ability to be able to market plants compatible with your company's products will become even more important.

Clonal Trees

To the biotechnologist a tree is a device for turning fresh air and sunlight into money with the help of some available land. To get the money you need to sell the tree's fruits or the latex extracted from its trunk, or fell it and sell its timber. This way of thinking is not exactly new. Take rubber trees for example. It was by using new "clones"—sets of genetically identical high-yielding trees—that vast estates were built up by foreign rubber companies in Malaysia and elsewhere some fifty or sixty years ago. Other species have not cooperated so easily with imperialist expansion. In the case of the oil palm, until recently new plants have had to be grown from seed. Today's palm oil seeds are the result of systematic hybridisation of thick-shelled "dura" mother palms and a pollen parent with shell-less fruit, the so-called pisifera type. The resulting "tenera" trees produce fruit with moderate shell thickness and an enhanced oil yield from the outer fleshy part of the fruit. Margarine, amongst other things, is made from palm oil, and Unilever grows millions of these trees for that purpose.

Since the formation of Unilever in 1929 through the merger of Lever Brothers and the Dutch firm Margarine Unie, it has grown into the largest and most broadly based consumer business in the world and is the fifth largest company outside the USA. A major concern in its commercial, financial and technological activities has always been the purchase of raw materials to make margarine at the cheapest possible price. One way of doing this has been to acquire the flexibility to move from one kind of oil to another as market prices have changed. Another has been to increase control over the supply of palm oils, by establishing the company's own plantations, or by wielding its enormous buying power in international markets to push down prices, and thus push down the returns to the exporting countries and local producers. One report on Unilever's activities relates how, in the plantation economies, whole communities have been shifted, labour has been transported from elsewhere, and a completely new system of values imposed on indigenous cultures. It has been suggested that, because of the company's overriding buying power, agricultural communities, transport systems, entire economies and governments have been reduced to a dependent state. Tariff policies, economic development programmes, trade treaties, political and economic concerns, and the nature and structure of the world trade in oils have all been influenced by the paramount interest of Unilever. It is a classic example of a multinational corporation that safeguards its sources of supply and attempts to impose its needs upon the agriculture, landscape, local economic infrastructure, and the culture of countries rendered desperate for the foreign exchange that a cash crop brings.

Continual research is crucially important to this process. As a former chairman of Unilever put it:

> The aim is always to enable us to switch from one oil or fat to another without any loss of quality. The texture, the keepability [sic], flavour and the nutritional value of our margarine must not be impaired. Nor must the colour, the lather or washing qualities of our soaps. Subject to that imperative we are trying at all times to put ourselves in a position to use less of the oils and fats which are in short supply and more of those which are easier to get. Our research has, therefore, been directed for years to making us more flexible, more able to use as many different oils and fats as possible for as many purposes as possible.

In 1968 a research group was established at the Central Research Laboratory of Unilever UK to develop a technique of tissue culture, for application to palms in Unilever's plantations. In March 1976 the first clonal palms were sent to Malaysia as small bare-root plants. They were planted the following year at Unipamol Kluang and began to bear fruit in November 1978. The implication of all this is that high yielding palm trees can be propagated asexually, copied if you wish, rather than created by the sexual process of pollination and the production of seed. The genetic roulette of sexual reproduction with its risk of varying characteristics is avoided, and the dissipation of valuable traits gathered together in one plant need no longer occur. Harvesting becomes easier because the trees within a given clone should be of more uniform size, the fruit ripen all at the same time and the oil be of more uniform composition.

Genetically identical trees, then, may keep the price of margarine down, or at least enable Unilever to maintain its commanding position as a food manufacturer. Palm oil trees have been cloned, and Unilever claims a thirty per cent increase in yields from the new trees, and work with the coconut palm is now going ahead. They also plan to sell their new varieties to other producers, including, no doubt, the nationalised plantations and cooperatives from whom they buy some of their raw materials. In one of the other three main central research facilities maintained by Unilever, at Vlaardingen in the Netherlands, scientists are working on analysing the genetic controls to oil production in the palm. The intention is to find out which genes control the production of which fats (lipids) in the oil and to move those genes around, so as to enhance the yields from particular varieties of palm. They want, in other words, to cross-breed at the molecular level. In the longer term, if this is successful, it would feed back to the production of clones, since specific genes could be inserted into the plant cells as they are growing in tissue culture, before developing into plantlets and thence into trees, tended by Malaysian labourers.

It is important to realise that the lipids in palm oil could be made in bacteria. In 1978, a rough estimate of the cost of doing this was £2000 per tonne. The implication is that anything that costs more than that when made by present-day processes of extraction from plant materials is worth *considering* for microbial production, even if in practice it turns out to be very difficult to achieve. Anything that costs less than £2000 per tonne, however, is best obtained by conventional means. Palm oil is currently selling at £400 per tonne. But as we have just seen, its price in world markets does depend upon the balance of power, both political and economic, between suppliers and producers. If the balance were to shift, as it has done with crude oil and petroleum refining, then the economics of making palm oil would change. In that sense work on the molecular genetics of palm oil could well be a long-term insurance policy against political changes affecting the price of the raw material. Were a bacterial process to be developed with a cheaper product then the exporting developing countries would lose another asset.

Unilever is not the only big company with an interest in trees. The

American timber concern Weyerhaeuser is also pioneering test-tube clonal trees. Weyerhaeuser is big. The family-controlled company owns 2.4 million hectares of forest in the United States and has harvesting rights on a further 3.3 million hectares in British Columbia, 405,000 hectares in eastern Canada and 607,000 hectares in the Far East. That makes it something like the world's largest timber products producer.

One option for a timber company is to keep moving. Forests are felled with the greatest speed, and the profitable timber ripped out with bulldozers and heavy machinery that simply reduce all the smaller trees and vegetation to pulp. That tends to be how chipboard gets made, through the wholesale gutting of tracts of forest, destroying insect life, plants, flowers, and (through the destruction of their habitat) birds and forest animals. When the available territory is vast and the authorities compliant (as is often the case with tropical forests) the juggernauts keep moving, and little reafforestation occurs. Even with the planting of new trees, ecological damage is substantial.

The most serious consequence of this arboreal strip-mining is that genetic resources are lost. Tropical forests are vast reservoirs of plant, insect and mammalian species. One estimate is that about forty per cent of the world's five to ten million species are to be found in these forests, which are disappearing rapidly. Half of them may be gone by the end of the century, taking with them virtually all of the species they used to harbour. That is a staggering prospect. These tropical plants and insects are not botanical or zoological curios. They are vital to the continuation of plant breeding and an immense source of useful substances and characteristics. The few species that we use as crop plants are continually improved by crossing with wild relatives, for instance the maize discovered in central Mexico which is expected to boost productivity in maize crops world-wide.

There are, also, unknown species of fruits, trees and shrubs which could be used for all kinds of purposes. The insects can be used for pest control, and butterflies particularly could prove an important source of antibiotics and anticancer compounds. Rather than grinding up thousands of farmed butterflies, which is a gruesome thought, one might grow their cells in culture and extract the benefits that grow. But, if the butterflies have gone, that option is closed.

Conservation then is essential, preferably by setting aside immense areas of forest as genetic reserves. This raises problems with landowners and timber companies, and for governments that want to sell forestry products for foreign exchange. Some timber companies do replant and treat the land as a tree farm, going for high-yield forestry based on research on planting density, optimal thinning rates and weed and insect control. But this only recreates a forest massively depleted in species. With this kind of scientific tree management it makes sense to select and use high-yield, fast-growing trees. Even with fast-growing strains the seedling to harvest cycle is about sixty years for a Douglas fir, and about forty years for a southern pine. To select the best examples, and to propagate them asexually, Weyerhaeuser scientists have also turned to tissue culture and the cloning of tree cells. At the moment progress is limited, although clones of some timber varieties are growing in test plots.

One problem with this form of selection and breeding is that, as we have said, it renders plantations or forests more genetically uniform. More and more trees over a given area come to resemble each other in many of their characteristics. Within the clones all the trees will be genetically identical though they will all look a little different, since that is what the term "clone" means—a *set* of like individuals, although the word is also used to refer to a *member* of such a set. Normally populations of organisms are diverse. Within a particular species some will have the capacity to resist certain diseases, say a blight caused by fungus, while others will not. It is the existence of such diversity that enables species to survive. If another outbreak of disease occurs, sensitive individuals will succumb, but those which are resistant survive and repopulate the areas where the sensitive plants have died out, although other species will be trying to colonise the same space. Through this kind of breeding programme, that collective resilience is bred out, and the population as a whole, say a tract of land put down to corn, becomes more vulnerable to pathogenic organisms or pests. In 1970 the US corn crop was attacked by a new form of blight-causing fungus, *Helminthosporium maydis,* to which the new strains of corn in cultivation were vulnerable. Fifteen per cent of the crop was lost at an estimated cost of between $500 million and $1000 million. Only swift reaction in producing a new and resistant corn variety prevented recurring agricultural disaster. We return to this issue later.

The conclusion I draw from all this is that the activities of big timber companies have the effect of reducing genetic diversity. But reduction of diversity makes species more vulnerable and has incalculable effects on the balance of nature. Business values are in direct conflict with ecological ones.

A Loose Box of New Animals

Biotechnology is just beginning to have an impact on the farm. We have seen how restructured plants may transform it. Other instances of possible developments might lie with what may happen to agricultural waste. Slurry—faeces—may be used increasingly to generate methane, although this hardly pays for itself at the moment. Plant wastes may also be fermented. More likely, though, is a change concerning which plants are grown, and who buys them. Some land will be turned over to the growing of fuel rather than food.

In the midst of all this activity it is worth considering what is likely to happen to farm animals, given that many people are likely to want to obtain much of their protein from the slaughterhouse, or more accurately, from the supermarket where they can forget about the ways in which animals are reared and killed for food. What do recombinant DNA research and applied genetics mean for the shape of things to come?

On one level, new feedstuffs which are made from bacteria are already available. ICI's Pruteen is just the first. Even with the high costs of getting into production, there are bound to be more. New vaccines for calves and piglets are with us now. Injectable animal growth hormone may soon be on sale. New antibiotics and monoclonal antibodies continue to be developed for veterinary use.

At another level, closest to the animals themselves, changes are under way in the methods of producing new animals with productivity in mind, rather than basic genetic shifts in form and anatomical structure. Artificial insemination is now widespread in dairy farming, and bull semen is an established commodity sold world-wide. Less well known is the analogous practice of embryo transfer (ET). Soon after conception, cow embryos can be removed from the female parent and frozen, so as to arrest further development without damage. In this state the embryos may be stored or sold to farmers who reimplant the embryo in another cow brought to the right stage in the oestrus cycle by hormone injections. The business is already worth some $25 million a year and is expanding rapidly. Basically the technique boosts the rate of reproduction of pedigree cows, just as artificial insemination has done for the best bulls. Supersires can now produce 100,000 offspring a year. For example up to sixty per cent of US dairy cattle are produced by artificial insemination. The figure for beef cattle stands at five per cent. The rate for embryo transfer is rather lower, with something like 20,000 pregnancies induced by ET in 1982, or 1 in 10,000 live calves born. It is estimated that very soon ten per cent of all stud bulls in the US will be born by ET. The techniques involved are not cheap. The hormone injection to induce superovulation costs around $2000, so ET is only applied to top-quality cows whose offspring can be sold for high prices.

The size of the returns on the initial investment in a pedigree animal has led to a new form of tax shelter, favoured by syndicates of small investors. A prime donor cow can be bought for between $20,-000 and $75,000 and can be made to yield say twelve high-grade calves a year, which can be sold for anywhere from $2500 to $100,000 in the market. Under US tax laws the expenses of ET, maintenance of the calves and the costs of marketing are allowable against tax, and this, together with a number of other concessions, means that the effective rate of tax on a substantial investment income can be very low. Needless to say the syndicates have to know what they are buying and how to sell the calves. Some of them don't.

To meet this kind of demand a number of small companies have been set up to market cow embryos, and cattle breeders all over the world are buying their products in order to raise the quality of their herds. Because of the expense this is only worth doing with pedigree cattle; less productive cattle, sheep and pigs are not yet worth the money involved. But other techniques are coming along which will expand the range of options. For example scientists are working on reimplanting twin cow embryos, trying to produce triplets in sheep, and attempting to raise litters in pigs from an average of thirteen to twenty or twenty-five. Another possibility is the sexing of embryos prior to implantation, which is now technically possible. Artificial insemination is another area in which this is continually being attempted, either by separating out male and female sperm, or by influencing the physical conditions of conception, so that only offspring of a given sex result.

More dramatically, pigs have been born after test-tube fertilisation (technically called *in vitro*), with the ova and sperm coming together outside the pig's body. Scientists working near Cambridge, England, have succeeded in dissociating the cells of developing sheep embryos so that each of the cells can be made to develop into an embryo in its own right, and thence into an adult sheep. One embryo can produce five sheep. The resulting progeny are genetically identical, having all started life as one individual. This is tricky stuff at the moment, requiring extremely delicate laboratory manipulation, but it looks as if it may be commercially applicable before long. Combined with superovulation it would further increase the productivity of animal breeding.

Looking to the longer term it may be possible to introduce new genes into animal cells at this stage in development. Experiments with mice described in chapter 1 involved the micro-injection of DNA into new fertilised germ cells, and this is also feasible with cells of agricultural animals. The problem at the moment is to relate particular DNA sequences to desirable traits. Since many genes control say, milk yield, it may simply be impractical to engineer higher milk yields in this way. This is a similar problem to that of gene transfer in plants. For the moment applied genetics is being used to increase animal productivity within the already established lines of the commercial market. As for the long-term future, we shall see.

Mechanisation

In the first chapter of this book I pointed out that new methods of making quinine may spell economic disaster for the people who grow or collect cinchona bark. We have just seen that biotechnical knowledge can be wielded as an economic force. Here I want to show that there is a similar political and economic dimension to the breeding and growing of tomatoes. Drinking a Bloody Mary or a gin and tonic may never be the same again.

New tomato varieties are being developed all the time. Throughout the world the annual production is about forty million tons, the bulk of which is made into tomato paste, soup and ketchup. In some parts of the world, such as California and the Midwest, tomato processing is big business. Consequently there is a great deal of interest in getting the most out of what characteristics tomatoes have by selective breeding. As I pointed out earlier, tomatoes can be grown from protoplasts, and this technique is already being used to develop new varieties.

What forces drive tomato breeding in particular directions? This question has been analysed by John Vandermeer, a biologist at the University of Michigan, and I draw on his work here. In essence his argument is that plant breeders were recruited to one side of the struggle between the big five tomato processors (Heinz, Campbell and Libby amongst them) and the agricultural labourers who pick the tomatoes, or who used to do so. Within this battle, mechanisation of harvesting was a stick with which the processors could beat the migrant labourers. To get mechanisation to work, new kinds of tomatoes were needed that could be made to ripen simultaneously and would withstand picking by machine and transportation in bulk. Plant breeders provided the new strains. This is the kind of project, with resultant losses to migrant workers and gains to food processors, growers and consumers, that plant breeders take on.

The mechanisation of the tomato harvest occurred in the late 1960s and early 1970s. While in California it took thirty years to go from one to ninety-five per cent of mechanical harvesting in cotton, it took only six years to achieve the same thing in tomatoes. Two important factors underlie this contrast. First, the strength of the United Farm Workers Union, whose campaign for badly needed higher wages and better conditions accelerated the trend to mechanisation on the part of the growers. Second, the research work on the tomato came some thirty years after that with cotton, and the structure and style of academic research had changed in the interim. By the late 1960s there was a great deal more experience of how to organise multifaceted programmes of targeted research, greater preparedness by government agencies to organise that kind of research and greater willingness on the part of members of a research establishment to join in.

Mechanisation requires uniformity, as the machines make only one pass through the field. All the fruit in a field must reach maturity at very nearly the same time. This means uniform transplanting of seedlings, uniform pest and weed control and response to fertilisers. That was not easy to achieve. Research on the biochemistry of ripening showed that if tomatoes were sprayed simultaneously with a special chemical called epethon, ripening could be artificially triggered.

A second line of research confronted the problems of rough handling. It was not difficult to produce tomatoes with thicker skins, but they turned out to ripen in a distressingly nonuniform way. Then they were susceptible to wilt, or they were too small. Finally breeders switched to a new shape, the elongated variety, and everything fell into place, although various biochemical parameters that affected the processing had to be sorted out. In addition the flow of tomatoes from the fields to the factories was changed by mechanisation, and the process for making concentrate had to be improved.

All this was taken on by study groups and think tanks of tomato researchers, many of them acting as consultants from a university base. Some of the money for all this came from interested corporations, but the bulk was from government sources.

Tomatoes in the Midwest are grown on land owned by small- or large-scale growers. The crop is presold to one of five large food companies, which strongly influence the market. They buy a set amount of tomatoes based on a known area under cultivation of between 5 and 200 acres and on an average of past yields from that land. The tomato plants are owned by the food processors. Most of them are handed out to the growers in the spring, and the growing crop is visited weekly by the food company agents. Harvesting runs on a schedule, as it does in the pea harvest in East Anglia, where representatives from Birds Eye, a Unilever company, instruct growers when to go into action.

Tomato harvesting is done by migrant workers, whose pay and living conditions are appalling. In 1969 they founded the Farm Labor Organising Committee (FLOC), which negotiated better contracts from the growers. However, the growers were soon unable to pay the union rates, because of their inability to get more from the processors for their tomatoes. Growers had either to change their line of business, to cut wages, or to opt for mechanisation. FLOC was clearly cutting its own throat.

Accordingly, it changed its strategy to one of a series of strikes in 1978, 1979 and 1980, directed against growers selling their crop to Libby or Campbell. One of the central demands of the strike was that FLOC be included in the annual contract negotiations between the canneries and growers. Its attitude to mechanised harvesting is generally favourable, but FLOC is demanding a retraining programme for those workers displaced from tomato harvesting.

Research by breeders was crucial to this continuing struggle. Vandermeer poses the question of what else plant scientists could have done in this situation. His answer was formulated by asking tomato pickers what could be done to improve their jobs. "Eliminate the stooping," "Plant tomatoes less densely, so that each whole plant can be scanned more quickly," "Stop spraying pesticides (so that children are not exposed to residues on the ground)," "Design a better container to receive hand-picked tomatoes," he was told. His question was then reformulated to ask how some of these goals could be realised without diminishing the number of jobs available or the price of labour. His speculative answer takes up the idea of decreasing pesticide use by developing systems of integrated pest management, where insect infestation is carefully monitored, some biological controls on insect proliferation introduced and some pesticides used. Within such a scheme some of the labour of harvesting would be transferred to pest control, with retrained workers undertaking some of the research and monitoring activity. The economics of all this are not worked out, though they don't seem implausible.

The real point is that these kinds of ideas could be worked up in a strategic way, by teams of researchers, just as the earlier programme of developing mechanisation was carried out. Various industrial interests, such as pesticide manufacturers and others, would be challenged by this, and, as Vandermeer shows, these are the concerns for which plant scientists tend to consult and for whose plans government funds are available. The problem, then, is to remove a research establishment held in thrall to the major corporations, to build support for a different set of scientific questions organised around the goal of satisfying, safe and secure employment rather than labour displacement and profit maximisation.

This, in a sense, is the problem that dedicated and energetic scientific servants of corporate capital represent. Much of their working lives has been devoted to building up the experience necessary to dovetail sophisticated research projects, in several disciplines, into strategic patterns dictated by industrial corporations. Corporate interest is not identical with public interest in many respects, as the tomato example shows, but the structure of material incentives and others in our society is such that most scientists never stop to think about why they accept the tasks they are given. It is this reorientation of the process of setting research priorities that foreknowledge of a "biotechnological revolution" allows us. Do we have the political will to bring it about, or will corporate interests come to be seen *as* the public interest?

These issues have surfaced elsewhere. For example in the late 1970s a radical Californian pressure group brought a suit against the University of California claiming that by allowing research on the mechanisation of harvesting to occur at the campus of Davis, in the heart of the Sacramento River Valley, the university was violating its charter which requires that research conducted there shall be in the public interest. Clearly such a challenge was going to fail, not least because it is so easy to argue that profitability in agriculture and the ending of backbreaking jobs like tomato-picking is straightforwardly a good thing. Also, in Michigan and Ohio, supporters of FLOC arguing for retraining found their arguments countered with blatant, unalloyed racism. "The Mexicans don't deserve it; they shouldn't be here anyway," and so on.

With that kind of opposition, to say nothing of the financial threat that it was thought to represent to faculty members and the university, the California Rural Legal Assistance lawsuit was on its way nowhere. But it did at least raise in public the question of why agricultural research is done, who is affected by it and whether universities have a responsibility to anticipate adverse social consequences of the technical changes that they make possible.

My view is that they do, even to the extent of questioning and trying to change the terms of reference of their contract research. The real danger of contractor-universities or research institutes is that if such a stance were maintained, getting research contracts would become next to impossible. Alternative options simply cannot be raised or considered if universities become merely the outworkers of industrial corporations. That is not to say the university or other fundamental research should not be applied or applicable—far from it—but that the goals of its application should, in the public interest, be widely debated and scrutinised.

Seeds: The New Key to Dependence

I doubt if many people in the developed world think of seeds as a strategic resource. So few of us nowadays plant our own food that seeds are a trivial product, tucked away in envelopes in hardware shops, as consumer goods for the keen gardener. Farmers and growers take a different view. For them seeds mark an important step in the annual cycle of planting and harvesting, investment and sale. But only the most informed and forward-looking can give much attention to the development of the seed trade as an industry, and to the long-term trends over perhaps half a century that influence the range of seeds on the market, their price and their effect on agriculture.

What is happening to the seed business is vitally important, and it forms the context for all the developments in plant science and technology considered in this chapter. Seeds are the starting point for a great deal of agriculture. They offer control over agricultural economies. They are the gateway to enormous international markets, if farmers can be persuaded or coerced into dependence on the commercial seed suppliers. Once seeds came in envelopes. For the amateur gardener they still do. But metaphorically speaking seeds have become the envelope for a whole package of chemical products on which modern agriculture is dependent. Seeds are a means to the enforcement of that dependence.

The development of plant breeding on an international scale, the revolutionising of present agriculture and the growth of an agribusiness sector together form a historical web of events that we need to understand for the meaning of plant biotechnology to be clear. Ultimately, making new plants—or making plant materials without plants—is about remaking the forms of dependence that bind growers and consumers of plants to the agribusiness corporations.

The green revolution of the 1960s has its roots in the 1930s. It takes about thirty years to grow an agricultural revolution. On that basis, the agricultural impact of plant biotechnology won't become clear until the twenty-first century. However, in the interim, the pace of change has already increased, bringing these effects that much closer. In the 1930s the first fruits of the new science of genetics, applied to plant breeding, came into agricultural production—hybrid corn with higher yields that built the US corn belt to its present strength. This achievement had an impact around the world.

In the Soviet Union it was enthusiastically endorsed by agricultural reformers wanting to create huge "grain factories" as a model of socialist production based on applied science. But the growing of corn is more difficult in the Russian climate, and the failure of these early enthusiasms in practice helped to compromise Mendelian genetics as an applicable science, as Soviet agricultural science spiralled into the battles over the political control of science of the Lysenkoist era.

Trofim Denisovitch Lysenko was a plant scientist born into a Ukrainian peasant family. Trained as a horticulturalist and plant scientist in the late 1920s he experimented with various techniques for modifying the germination of wheat. His aim was to allow winter wheat to be planted in the spring, when conditions were less severe, and yet ripen quickly. The apparent success of his technique led him to claim that here was a way in which new varieties could be created rapidly, without recourse to the lengthy breeding programme and laboratory science that the geneticists claimed were necessary. Under the tremendous pressures of the Stalinist drive for industrialisation, in which increasing agricultural productivity was a crucial element, Lysenko's claims for his theories, and his attack on Mendelian genetics as useless, reactionary and inconsistent with Marxism, produced a major political storm. By the late 1930s, Lysenko and his supporters had gained ascendancy over the Mendelians. By 1948, his position was such that Mendelism was actually banned from Soviet biology, to the consternation of European and American biologists who found Mendelian genetics increasingly useful and who thought Lysenko's ideas nonsense. Although this interdiction only held for a few years, until

Lysenkoist theories were discredited, it set back fundamental and applied genetics in the Soviet Union for a long period.

In America the success of targeted plant-breeding programmes made an impact on the foundations. In the early 1940s, the Rockefeller Foundation established a centre for work on wheat and maize in Mexico which became the world-renowned centre for the improvement of wheat and maize (CIMMYT). It would perhaps be wrong to see this as disinterested philanthropy only, although in what sense it was "interested" or conceived as a strategic move is not clear. One possibility is that it followed from major irrigation projects in Mexico in the 1930s. Another is that the surplus of labour there made it an attractive location for plant breeding, which can be a labour-intensive activity. Yet another is the proximity to traditional cereal varieties in cultivation in traditional agriculture. A more plausible and significant reason is that American foreign policymakers in the 1940s perceived Mexico as a strategically important country, bordering on the United States, that needed to be modernised without a revolution and linked harmoniously to the expanding US postwar economy. The official version is that the Rockefeller Foundation was persuaded by the progressive American agriculturalist Henry Wallace to do something about world hunger.

As the programme at CIMMYT developed it came to be seen as a model of what private philanthropy or government-sponsored research could do to modernise agriculture in politically unstable, less developed countries, where food and land shortages could produce unbearable and unmanageable political tensions. Above all it represented "managed" social reform through strategic technical change, the hallmark of Rockefeller philanthropy. With a more "efficient" agricultural base would come political stability and an increase in international trade, particularly, as it turned out, for the products like fertilisers, pesticides, irrigation pumps, farm machinery and fuel required to make the new agriculture work and the new plants realise their potential yields.

The Rockefeller initiative was followed by other foundations, such as the Ford Foundation, which helped to establish an International Rice Research Institute, and the Kellogg Foundation. As the costs mounted the funding for the plant-breeding stations and related activities was transferred to national governments and international agencies, such as the World Bank and the United Nations Food and Agriculture Organisation.

By the mid 1960s new varieties of wheat and maize were available.

Under the right conditions these could produce massive increases of
five or ten times the yield. Huge amounts of aid money were injected,
to help poor farmers or, indeed, poor countries buy the seeds, fertilisers, pesticides and fuel needed. The acreage of these high-yielding
varieties increased dramatically, as shown below:

	1965	1973
Wheat (in hectares) (Mexico, India, Turkey and Pakistan)	10,000	17,000,000
Rice (in hectares) (Taiwan, Philippines, Sri Lanka and India)	49,000	16,000,000

The positive economic effects of this rapid diffusion of new varieties
have been considerable. Cereal yields in Turkey since 1970 have doubled to eighteen million tons per annum, yet the Turkish economy is
massively dependent on international loans, and thousands of Turks
work outside their country. In Mexico wheat yields have leapt from
twelve to fifty bushels per acre, yet the Mexican economy is in continual crisis, and its debts a continuing "problem" for international bankers. By 1972–73 high-yielding varieties were contributing a billion
dollars a year to Asian cereal harvesters.

Supporters of the green revolution understandably continue to
stress gains like these, although since the 1970s they have been forced
to accept that the need for massive amounts of fertiliser and other
chemicals in order to produce the higher yields is a major economic
flaw in the green revolution strategy, and that its unfortunate social
and agricultural effects cannot be ignored.

The new varieties demand an increase in the amount of capital
necessary to farm, and in the amount of *wage*-labour. To grow many
of the new varieties, one needs irrigated land, fertiliser, insecticides
and fuel. To buy these, money or credit is needed, either from a bank
or from the government. Richer Third World farmers can muster the
resources and are allowed to take the risks. The vast majority of small
subsistence farmers can't, and they are being forced off their land
because of higher rents or because they can't match the competition
from the bigger producers. So they become landless labourers or take
off for the cities. Those remaining in the countryside have shown
themselves increasingly ready to bargain for higher wages which has
in turn encouraged a trend to mechanisation.

It must have been clear that this would be the case from the early
days. Some of the architects of the green revolution, however, saw this
as positively healthy, as the discipline of the market made itself felt on

small producers, used to barter, to feudal obligations and traditional, symbolic patterns of cooperation and exchange. What they seem not to have foreseen is that the provision of government subsidies and the establishment of credit arrangements provided very effective means of economic, political and social control. The control of cash for farmers, exercised at the village level by the local hierarchy, determined who, if anyone, was to benefit from the new agriculture.

In agricultural terms the new varieties have been a very mixed bag. In some cases yields have increased dramatically with the right inputs of chemicals, adequate labour, and soil and climate conditions that match those in the region where the new varieties were developed. In others the yields have been below the claims made for them by the seed corporations. Sometimes the new crops have attracted new pests, like the locusts that decimate rice harvests in Pakistan. In particular some have proved less resistant to known plant diseases, and because of the vast acres under cultivation of the new varieties the impact of these diseases has been much greater.

As initial enthusiasm for the new varieties grows, the traditional varieties in cultivation are abandoned and their seed is no longer collected. Not only are crops as a whole becoming more vulnerable to pests and diseases because of increasing genetic uniformity, the established local varieties, from which new plants could be bred or to which production could revert if the new ones fail to offer a significant improvement, are disappearing. Moreover many of the new varieties are designed for different systems of agriculture, which do not allow for intercropping, for example growing one plant between the rows of another crop. Some peasant farms grow vegetables between their cereals, keeping the former for their own consumption and selling the latter for profit. With the high yielding varieties this is no longer possible, and a valuable source of food is lost.

One view of all this is that these are side-effects and unintended consequences of agricultural modernisation. The implication is that they can be managed, that more enlightened administration will alleviate the problems of conserving traditional varieties or making credit more widely available or concentrate more research on tropical vegetables.

Another view is that the "modernisation," around which the green revolution was organised, was intended to stimulate international trade on terms that exploit less developed countries. On this view one can see the initial philanthropy as a move that revealed the economic possibilities of investing in plant improvement. Through the capitalisation of peasant agriculture came increased demand for agricultural capital goods. For developing countries without the money to buy the fertilisers and seeds, international loans were set up, repayable as economies expanded. But as the costs of fertilisers, pesticides and fuel went up in the early 1970s, developing countries have found themselves committed to agricultural systems that they can no longer afford to run. To sustain the production of food some have been tempted or forced into negotiating further development loans, in order to keep buying the materials their farmers need.

On this view the green revolution was a means, initially unsought perhaps but certainly not unrecognised now, of strengthening economic dependence, by building up the need for new seeds. Following this first phase of the green revolution, the second has seen the entry of multinational oil, chemical and pharmaceutical corporations into the seed industry, since the markets are now global and the opportunities for market domination enhanced by changes in patent legislation.

One sign of serious investment activity in an industrial field is the production of technical or commercial forecasts offered to potential investors at incredible prices. In 1978 an American consulting firm, L. W. Teweles & Co., offered for sale its report, *The Global Seed Study*, at $25,000 a copy. In a letter to prospective buyers, quoted by Pat Mooney in his book on the political economy of seeds, the author wrote: "In the last ten years, at least thirty seed companies with sales of $5 million or more have been acquired by large, nonseed multinational corporate enterprise. At least eleven more such mergers are believed to be under discussion." Mooney describes this as an underestimate, and cites evidence such as that, in one week, Rank Hovis McDougall in the UK bought up eighty-four companies, when legislative changes meant that patentlike protection and royalty payments, so-called plant breeders rights, became available on new varieties. Similarly in the United States, following the passage of the Plant Variety Protection Act in 1970, which massively extended patent protection to sexually reproducing plants, the American Seed Trade Association devoted half its annual meetings to a special symposium called "How to Sell Your Seed Company."

Mooney and others have argued that this tremendous wave of acquisition, with some of the world's largest companies like ITT, Royal Dutch Shell, Sandoz, Ciba-Geigy and Union Carbide buying up small and large seed concerns, was catalysed by the dramatic extension of patentability to virtually all new varieties in the 1960s and 1970s. In Europe the enabling legislation was passed some ten years before that in the USA. The first plant patents in the USA were made possible by the Plant Patent Act of 1930 that covered asexual reproduction of varieties of plants that could be pirated by a competitor simply by taking a cutting. Since that time the arguments have raged between the seed or plant stock producers, arguing for patent protection as a commercial necessity and a spur to innovation, and consumers of seeds, usually farmers, but sometimes food companies, who argue that patents will allow prices to be raised and will have unfortunate consequences on plant breeding and on the conservation of genetic resources. Thus in 1970 one influential lobby against the US Plant Variety Protection Act was that of the soup manufacturers, like Heinz and Campbell. They were able to force an amendment to the Act. Under this amendment, it was impossible to patent new varieties of tomatoes, celery, carrots, cucumbers, okra and peppers. The situation changed again in 1980, when this restriction was removed.

In Europe the position is similar, although it is linked to legislation that requires the registration of new varieties which must live up to their trade description and offer an improvement over existing ones, and permits only the cultivation of varieties so registered. The effect is on the one hand to make it difficult to get new products into the seed catalogue, but on the other to decrease, quite dramatically, the number of varieties in cultivation and to allow domination of the seed markets by powerful concerns with the resources to get their products patented under this system. The picture then seems to be that a small number of multinational firms dominate the sales of seeds in at least these six important crops. With this in mind, [in 1980] the UN Food and Agriculture Organisation stated:

> In fact, the growing concentration of plant breeding in the private sector has already demonstrated some negative effects. Among these are to be noted, for example, the increased cost of development programmes linked to increased cost of seed and related inputs, the use of marketing techniques inappropriate in developing countries, which have led to grossly unbalanced agricultural inputs, and crops on top-grade food-producing agricultural land owned by multinational companies and intended for foreign markets.
>
> Furthermore, since the germplasm of most of the world's important crops originates in developing countries, while most plant breeding, particularly sophisticated private sector production of new varieties, is conducted in developed countries, in an increasing number of cases the developing countries have been required to pay royalties for varieties, the germplasm of which originated within their borders.

Other critics of present trends in the seed industry, or as one consulting firm encourages us to call it, the "genetics supply industry," have also pointed to the shutting down of free communication and the exchange of breeding material. Gary Fowler, representing the National Sharecroppers Fund at the Senate Subcommittee hearings on amendments to the Plant Variety Protection Act in 1980, noted:

> Since 1972, not a single agribusiness breeder has published descriptions of breeding schemes or techniques for their new varieties in *Hort Science*'s "Cultivar and Germplasm Releases" Section, the most popular outlet for such information among university and government breeders.

Another striking indication of how aggressive commercial behaviour oriented to patenting is likely to shut down communication between researchers came up in autumn 1982. In April 1982, the US Patent and Trademark Office had issued a patent on a technique for the accelerated production of new hybrid strains of plants and the rapid commercial production of seed for such hybrids. The patent, No. 4,326,358, was awarded to the American seed firm Agrigenetics of Denver, Colorado, which has been built up into a business with a turnover of $100 million through the takeover of small seed businesses by its chief executive, David Padwa. Agrigenetics has also moved into the field of plant biotechnology and has several distinguished academics on its payroll as consultants.

The granting of this patent took plant breeders in Britain by surprise, for in their view the technique concerned is both well known and in frequent use. If the patent is maintained, after indirect procedural challenges and possible litigation, then plant breeders will have to pay royalties to Agrigenetics if they wish to propagate hybrids in this way. In a letter to *Nature* in August 1982, Professor Neil Innes, a member of staff at the National Vegetable Research Station in Britain, aired these angry thoughts publicly. He documented his claim that many elements of the procedure have been well known for years and that the crucial feature, the micropropagation of selected plants in tissue culture, had been discussed in the technical literature in 1978. Accordingly Professor Innes, as chairman of the British Association of Plant Breeders, asked the US Patent and Trademark Office to re-examine the application and to withdraw it on the grounds that they had overlooked its dependence on already published ideas, that it was indeed "obvious" and no advantage beyond what the patent law calls "prior art." A second Agrigenetics patent, on the application of this technique to a particular species, is also under application.

The sense of outrage at this move by Agrigenetics is plain in Innes' letter. Clearly he feels that a commercial concern has provocatively and brazenly grabbed a basic technique for itself. It is all very reminiscent of the Cohen-Boyer patent, and the controversy over the patenting of monoclonal antibodies, except that in this case the organisation seeking property rights over a fundamental technique is a commercial company and not a university. It will be interesting to see how the struggle develops and whether any large seed producers enter the lists against Agrigenetics. It may be that they will accept this precedent as one that they could in future use for themselves.

Not Yet Concluded

We find ourselves, I think, in a crucial period in the history of agriculture. Decisions that are made will have a major impact on the form, nature and output of agriculture for millennia. Processes that are now under way, such as the destruction of tropical rain forests, could have catastrophic effects on the price and availability of food in twenty or thirty years time. Without a wide range of wild species from which to select new breeding material, plant scientists will fall back in the constant struggle to keep plants at their present high levels of productivity, and to sustain their resistance against pests and disease.

Not surprisingly, some commercial organisations realise this, and are involved in international gene conservation programmes. Pioneer Hi-Brid, one of the world's largest seed companies with major interests in wheat and maize, provides a financial support for gene conservation —that is, the collection of species—carried out by CIMMYT in Mexico. United Food (formerly United Fruit Company) is estimated to have three quarters of the world's banana species in private collection. This constitutes a massive advantage over competitors.

At the same time international organisations funded by the United Nations are engaged in the collection of seeds and plants. Many thousands of species are being stockpiled in this way, although it is clear that the rate at which species disappear through the destruction of forests for timber and agriculture, urbanisation, the tendency to use new varieties and to ignore the traditional ones, and the trend to monoculture, will outstrip the rate at which species are being conserved.

But even if enough resources were devoted to halting the effects of "genetic erosion," there would still be a problem with access to and use of the materials in plant gene banks. It is not just that the world's "seed corn"—and I use that term metaphorically—is being used up; an ever greater proportion of that which remains is being used by national organisations and companies from the developed world. Their position as the inventors and sellers of new varieties is protected in many countries by legislation that confers patentlike protection on the owners of new species. Critics of such legislation in the USA, Canada, Britain and Australia have argued that it has the effect of further concentrating power in the hands of influential producers, who are able to turn the markets for particular plants into an oligopoly, dominated by a few semicompetitors. Its supporters, amongst them the lobby to commercial plant breeders, ASSINSEL, claim that the legislation is a stimulus to innovation. ASSINSEL was so perturbed by Mooney's book *Seeds of the Earth: Public or Private Resource?* that it issued a line-by-line critique of the book to all its members. When I wrote to ASSINSEL asking for a copy, my request was declined. I was sent instead their standard publicity brochure, which is long on rhetoric and short on analysis of the issues that Mooney raises.

As things stand, Third World countries do find themselves in the position of buying plants that have developed from their indigenous resources and which are protected by varietal legislation. It seems to me unlikely that they could ever form a cartel in plant genes, as OPEC did with oil, but some way must be found to break the dependence of these countries on the "genetic supply industry," and to get their assets properly valued, rather than stripped.

Such is the background for plant biotechnology. It is part of a power struggle between the breeders and users of plants, to which biotechnology is recruited either to enhance the control of industrial seed producers over agricultural consumers, or to maintain the low financial value of plants indigenous to some less developed countries. The new techniques of plant gene manipulation are being created to service that process, leaving the balance of power unchanged. It is possible that the overall result will be more food, but only at the price of immense risks in the long term as the number of species shrinks, and of continuing enormous imbalances in food's global distribution. And yet the potential of food biotechnology is that it could conceivably feed the multitudes. It is a stark example of the fundamental question for biotechnology: who shall set the priorities and who shall own the knowledge?

6

Pipeline into the Future: Chemicals and Energy

Rationalisation

When you look at carefully shot photographs of new chemical plants you would never guess that they smell. Not unless you were unlucky enough to live next to one. Nor would it come immediately to mind that a process plant wears out, leaks, breaks down and occasionally explodes. Moving millions of tons of highly reactive, corrosive and toxic liquids and gases through miles of pipes at very high temperatures and pressures has to be difficult. Just as photographs can be made to tell an incomplete story, so too the present rhetoric from the top of the chemical industry is misleading and euphemistic. Increasingly these days, leading executives have been gathering together to talk of the crisis in their industry and the need for "rationalisation." Some of it is brave talk, no doubt, carefully crafted to impress the competition and to reassure investors and stock market analysts with its air of realism and determination to survive.

However, "rationalisation" means among other things plant closures, redundancy, termination of contracts, wage cuts and changes in working practices. It means a continual state of struggle over who and what survives. The chemical industry world-wide is now riven by that kind of conflict; and so, of course, are many other sectors of production. It is likely to last for a decade or more. This is the context for an awakening interest in biotechnology as a road to salvation. The prospect of keeping all that plant in production, all those tankers on the motorways and all that money in circulation is beginning to look rather daunting. Biotechnology might, in the longer term, keep the show on the road, and it might give us a cleaner, safer more controllable, decentralised chemical industry.

It would be hard to overestimate the economic, financial, political and industrial importance of the chemical industry. ICI is the third largest employer in Britain. When the company made a quarterly loss for the first time, it was a major item on the national news and a source of anxiety in the stock market. Its pension fund alone is a major investor and can have a powerful effect on industrial takeovers. For some countries, to have a chemical industry is a matter of national prestige, like having an airline. Equally the products of the chemical industry may seem very prosaic, but they are utterly essential to urban, industrial life as we know it. These include plastics for shoes, kitchenware, cable insulation and containers, solvents like paint strippers, nail polish remover and dry cleaning fluid, paints, resins, artificial fibres for clothes, plasticisers and flavourings for our food, fertilisers, pesticides, pharmaceuticals, anaesthetics and antiseptics. Chemical products are both the symbol and substance of modernity.

In the 1950s, the annual growth rate of the chemical industry in Britain, West Germany, the USA and France was around twenty-five per cent. In the following decade the pace slackened to a mere fifteen to twenty per cent. In the early 1980s we are down to a few per cent, and for some companies to negative growth rates. World sales of petrochemicals are now around $690 billion per year, and in terms of growth, profitability, rate of innovation and technological performance the industry during the postwar period is an example of remarkable success.

Four factors underlay this growth, and all of them are now ceasing to operate. Firstly, there is the price of oil. The immensity of world oil reserves, the enormous excess of supply over demand and the power of the developed, oil-consuming nations over most of the producers has meant that for much of the postwar period oil has been very cheap. One of its many "fractions," naphtha, has been a favoured starting material for the chemical industry throughout the postwar period. (A "fraction" is a component that can be distilled or separated from the crude oil.) Before then, naphtha was regarded as waste and dumped back down oil wells when they ceased producing. One could hardly find a more telling illustration of how the status of waste materials can be transformed.

Since the early 1970s, when the oil-producing nations at last escaped from the clutches of their consumers and began to bid up the price of their form of energy, the situation has changed dramatically. The price of oil has gone up by a factor of eight between 1960 and 1980. How price increases of this size have affected the chemical industry we consider in a moment.

The second factor facilitating growth was economies of scale. With current technology (and within certain limits) the cost of producing the important chemical intermediary ethylene goes down significantly as the size of plant producing it goes up. By building bigger and bigger plants, the economic returns could be increased as long as the market continued to grow, as long, in other words, as one could continue to sell bigger and bigger amounts of chemicals.

This trend, extending through the 1950s and 1960s, was due to two further factors, substitution and general economic expansion. More and more plastics and petrochemical-based products replaced traditional materials. Higher wages, greater corporate purchasing and increased public expenditure meant that the demand for the products of the chemical industry continued to increase. All these influences are now no longer operative or operate in different ways, and adversely affect the chemical industry in Western Europe and the United States. Big energy-intensive plants built on optimistic projections of demand no longer offer the kinds of return their proud planners had in mind. The installed capacity is now too great to service very slowly growing markets and generate the expected profits. Overcapacity and overproduction are exacerbated by the effects of a shared desperation to sell. Individual, opportunist acts of price-cutting serve to keep prices for petrochemicals at levels that finance directors know to be dangerous yet are powerless to change. Added to that is growing competition from industries established by or in newly industrialising nations or in countries whose energy reserves are only now being fully opened up.

The interesting thing is that all these growth factors are said to have been taken for granted for too long. This is quite remarkable. A global, technologically sophisticated and wealthy industry like that based on petrochemicals must indulge all the time in an orgy of forecasting, simulation, scenario-building and economic modelling. One would think that the overall vulnerability to a rise in the price of oil, to the law of diminishing returns, to market saturation and a downturn in the world economy would have been foreseen much earlier. Maybe it was, but at that time sheer size stood in the way of flexibility. Maybe Cassandras have no career prospects.

All this can be fitted into several neat models of industrial development, patterns that order the past and the present, even if they don't clarify the future. One such is a wave theory of growth, put forward by an executive from the Dow Chemical Company in Europe. Since the war there have been two waves of expansion, and a third is under way. In the early 1940s, chemical companies, stimulated by the war, began to develop new refining processes and the manufacture of thermoplastics in bulk. This occurred on traditional sites, where the companies had been founded. These were not always well placed for easy access, distribution or mass production.

By the early 1960s, a new generation of much bigger chemical plants was coming on stream in new places, closer to harbours and motorways and a little further away from urban areas, in order to moderate the pollution. The older sites were turned over to speciality chemicals. This period saw relocation to Texas, Louisiana and Tennessee, in the US, and Rotterdam, Marseilles and northern Spain in Europe.

Now a new wave is beginning, close to politically secure sources of feedstocks, and not too far from the markets. Alberta in Canada, Mexico and Central America, the Gulf of Iran and Indonesia are all places where this kind of significant investment is going on, although the second-wave sites are still important. In some of the new places labour is cheap and unorganised, and environmental controls are embryonic, lax or nonexistent. One implication of all this is that the way the world market will be made up regionally will change in the next twenty years. Even so, Europe is expected to constitute the largest market for petrochemicals, though its importance relative to the United States or Japan will diminish.

So far, then, we have had two waves of development that brought into existence or consolidated perhaps twenty or thirty multinational concerns and a hundred or more smaller firms. They operate vast energy-consuming plants that endlessly elaborate the structural permutations of hydrocarbon molecules, to churn out the materials that are basic to modern consumer society. Those concerns now feel themselves to be under great strain because of the rise in the price of energy and the related rise in the price of the starting materials. Already boards of directors are ordering the shedding of old plant and jobs in an attempt to survive. In a period of recession, with the prospect of bankruptcy all too imminent, the role of R&D is to help companies stay in the game and, if possible, find a route to new processes, products and markets that will allow re-expansion in the 1990s or the next century. Research is now even more crucial to staying in business as a chemical producer. Be that as it may, only well-heeled companies can spend the money. That, as it happens, has opened up opportunities for those who make it their business to sell technical forecasts, up-to-the-minute information and research ideas. Biotechnology has brought on a major landslide of extremely expensive surveys, newsletters and conferences. More often than not they are the glib talking to the desperate, or the underexperienced overselling to the overpaid.

How then does the energy squeeze work, and what could be done about it? Analyses of how much energy was consumed by the chemical industry, either as a fuel, to get the required temperatures and pressures, to generate steam and so on, or as a feedstock show the relative dependence on various materials and the size of that dependence measured as energy, not as money. Feedstock consumption is about nine per cent of the total US fuel consumption. Less than one might expect perhaps. In money terms, one can say that the UK industry, in 1979, spent about £830 million (six per cent of turnover) on energy. Roughly the same amount (£800 million) went on feedstocks. If you consider that roughly each year the chemical industry as a whole has either to find an extra £600 million or reduce its energy bill by half just to stay where it is, then the impact on profits is pretty clear. The alternative of passing increased costs to the consumer is not possible at a time when demand for its products is falling. In this situation one rather obvious thing to do is to try to use less energy. Chemical plants are now designed with that factor firmly in mind.

Another is to balance the sales value of particular kinds of chemicals against the energy required to produce them. Broadly speaking, organic chemicals, inorganic chemicals, dyestuffs and fertilisers are costly in energy for the revenue they produce, whereas pharmaceuticals, cosmetics, paints, soaps and speciality chemicals produce more revenue per unit of energy used to produce them. Now, one cannot immediately translate this contrast into relative rates of profit. A substance could require a lot of energy to make and still be profitable to produce. But it is a guide.

Could Biotechnology Be the Answer?

Biotechnology fits in here in several ways. First, one strategy for chemical companies is to concentrate more effort on speciality chemicals, where the rate of profit can be higher and the competition less. Medical products fall into this category, including pharmaceuticals and diagnostic reagents. Most large chemical firms have a medical division or are beginning to assemble one, as biotechnology opens up a whole new range of specialised, expensive molecules to sell. Second, biotechnology offers new feedstocks, starting with complex hydrocarbons like cellulose or sugar and breaking them down into simpler fractions, which can then be reprocessed into larger molecules. Third, biotechnology opens up the prospect of much lower operating temperatures and pressures, because living systems don't need and can't stand the extremes of modern chemical engineering. If living systems or isolated biological molecules can promote particular reactions, the process would use less energy than is now required for the same purpose. The trick, of course, is to make all these things work on an economic basis. At the moment the situation is very uncertain, not least because the amount of capital you have to invest to develop, say, a new feedstock is vast, and there are no guarantees that at the end of the development process the new material will in fact be economically worthwhile.

One strategic option is to reconstitute the chemical industry so that it can begin with coal. World reserves of coal are thought to be substantially larger than those of oil and gas. Coal was once used as a feedstock, only to be replaced by the cheap oil of the postwar period. It is made of the right materials—carbon, hydrogen, nitrogen and oxygen—to form the basis of much more complex molecules containing other useful elements like chlorine, iodine, bromine, sulphur and boron. It also has some drawbacks. It is solid, which means that you have to work on it to get it to flow through a continuously operating plant. It contains a lot of impurities that form gritty, abrasive, sticky ash that can clog everything up. Then its carbon chains are not in the best configuration for chemical processing, but they can be restructured if one is prepared to spend the energy. So it makes sense to tear down coal to a mixture of hydrogen and carbon monoxide called synthesis gas and then go from there, via an intermediate, like methanol, a simple alcohol. That too takes energy.

However, some companies—ICI, the German firm Lurgi and Mitsubishi in Japan amongst them—are working on turning synthesis gas into methanol, which can be used as a fuel or feedstock. It is said that only one company starts with coal-based synthesis gas, African Explosives and Chemical Industry in South Africa; the others make it from natural gas. Methanol looks like a chemical with a big future. At the moment it tends to be used to make speciality chemicals or products like acetic acid, but as new uses appear, projections are for a 400 per cent increase in demand by the year 2000, with slightly more than half of that going on new uses. It could serve as a feedstock for making chemical intermediates like ethylene and styrene, or as a fuel in combustion engines or fuel cells, or as the medium that will turn powdered coal into a transportable, pumpable slurry, or as part of the nourishment for bacteria producing single-cell protein. All this has a fairly traditional feel to it. Fossil fuels are still broken up into their constituent parts in extreme physical conditions. However, it does represent a major innovative step for both chemical companies looking for cheap raw materials and energy corporations looking for a new liquid fuel.

A more radical step is to move into fermentation and plug in directly to the degradative and synthetic potential of living organisms. That is taking a different route using the biotech highway, with "biomass" (biological materials that can be broken down into simple hydrocarbons) as a source of energy or as the feedstock, or one can turn to fermentation processes in which living organisms themselves produce industrial chemicals. Nobody expects the shift in biotechnology to occur overnight, but it is the wave of the future. Already some corporations are taking considerable steps in this direction. One third of the £200 million that ICI spends each year on R&D goes to the life sciences, much of that on pharmaceuticals, but a substantial amount must still go to new kinds of biotechnology. Du Pont, an American chemical company big enough to buy its own multinational oil company, Conoco, is making the plant sciences a major element in its evolution away from being a corporation that spun its profits out of nylon.

A Ferment in the Industry

Trying to chart the different options for the chemical industry is a bewildering task. Several things complicate the picture. First, there are the blurred boundaries between fuels and chemicals. Crude oil is separated into its fractions by carefully controlled heating in refineries. Oil companies have, therefore, diversified into chemicals over the years. Equally, chemical companies are beginning to buy into the energy sector. ICI has a stake in the North Sea and is drilling for oil around the world. Although the fuel business is different from chemicals, they overlap and interpenetrate. Biotechnology could continue the ambiguity as the products of fermenting organic waste or crushed oil seeds could be used as a fuel or a feedstock.

Second, there is a distinction to be made between fermentation processes, which depend on the metabolic processes of micro-organisms, like yeasts, and nonfermentation processes, like crushing oil from a seed. The latter uses biological materials, but not living processes.

Then there is the question of whether the starting materials are waste or whether they are deliberately grown to service a particular process. For example, you could take the wastes from forestry, which contain various kinds of cellulose and hydrocarbon chains, and turn them into alcohol. Or one could plant fast-growing trees specifically to provide a chemical feedstock or a fuel. One example is the fermentation of starch from plants like cassava in Brazil, in order to make the gasoline-alcohol mixture "gasohol." Ventures of this kind have their own advantages, such as savings on imported fossil fuels, and problems, such as changes in land use threatening food supplies. When the same process is run with waste materials, a different set of questions is raised such as whether such recycling and pollution control should be actively promoted and, if so, by whom and with what aims in mind.

To try and cope with this complexity I am going to take things in three passes, firstly the biotechnological production of industrial chemicals, principally by fermentation; secondly the development of "green" energy sources; and finally the processing of waste to produce chemical foodstocks, food and more easily disposable or controllable waste. This means that I shall have to discuss the same processes several times in a few cases, but this seems the clearest way to get through the maze of issues.

Fermenting Chemicals

Enzymes were discovered in the late nineteenth century by microbiologists who found that purified extracts from living cells would still catalyse—speed up—specific reactions, like the fermentation of sugar to alcohol. Nowadays the number of enzymes known to operate in biological systems is vast, several thousand at least, and a growing number are produced on an industrial scale as products in their own right with specific uses. Indeed one company, Novo Industri in Denmark (which we have already considered as an insulin manufacturer), gets most of its business from the sale of enzymes, particularly for detergents. Novo and the Dutch firm Gist Brocades control sixty per cent of the world market.

It is worth bearing in mind what enzymes are, namely large protein molecules, which act as highly specific biological catalysts. Their three-dimensional structure is such that they will bind to a specific substrate as a template to a mould. The part of the enzyme involved in this act of union is called the "active site." The substrate is the substance transformed in the chemical reaction that the enzyme speeds up. The enzyme comes out of the encounter unchanged. Since enzymes are large polypeptides (strings of amino acids) their chemical synthesis is not really feasible, as so many steps are involved and in each one a significant amount of material is lost, so you end up with precious little enzyme. Instead they are extracted, usually from bacteria, sometimes from other cells.

Some examples of enzymes: papain, used as a meat tenderiser, comes from the papaya fruit. Bacterial proteases, enzymes that can break up protein molecules, are used in biological detergents. Glucamylase, alpha-amylase and glucose isomerase are all used in turning corn starch into a high fructose corn syrup, which is increasingly being used as a sweetener in soft drinks in the United States. In Europe sugar-beet farmers are protected against the import of competing products by the tariff barriers of the EEC, so the use of this sweetener is unlikely to expand in Europe unless agricultural policy can be modified so as to encourage the aggregate use of sugars of all kinds. Rennin, an enzyme normally extracted from the fourth stomach of calves or cows, is used in cheese-making. It is an expensive little number: twenty-six tons of rennin cost $64 million in 1980. That works out at around $1200 a pound, which is rather more than the price of even the best cheese. It has recently been made by Collaborative Research, working under contract to Dow Chemical, by transferring the rennin genes into bacteria. Plans are afoot to get rennin made in yeast in the same way.

Genetic engineering has something to offer here by gearing up the production of specific enzymes from bacteria. One way to do this is to insert multiple copies of the gene specifying the enzyme into the bacterium. Another way is to splice in regulatory genes called promoters that enhance the cellular output. Another is to trick the microorganism into exporting more enzymes through its external membrane, making the extraction process easier. Japanese scientists have already increased the yield of the enzyme alpha-amylase from *Bacillus subtilis* by 200 times. The production of ligases, the repair enzymes essential to gene-splicing, has been increased 500-fold. Yet another possibility is to use bacteria that like living in very hot liquids, the so-called thermophilic bacteria, which have evolved to exploit rather uncomfortable niches like sulphurous springs. Enzymes from such bacteria can withstand higher temperatures without breaking apart, and the reactions they catalyse go at a faster rate. So if one could transfer the genes for alpha-amylase into a heat-loving bug, then the more rapid conversion of starch to glucose might be possible. Another advantage of using bugs like these is that the fermenter in which they grow doesn't have to be cooled so much, which saves energy that would otherwise go in artificial cooling. There is also a growing market for enzymes in making jams and alcoholic drinks, and in making certain pharmaceuticals.

There are two ways in which enzymatic processes might be cheapened. The first is to bring together several reactions within the confines of a single cell, by splicing the genes associated with each enzyme into chosen micro-organisms. That way you only need one fermenter. Each of the billions of cells in it would carry out the various reactions in sequence, fermenting the substrate to the desired end-product, with a saving in time and capital. This is to use the workings of cells to exploit their potential as reprogrammable, integrated systems. Another possibility is to go in exactly the opposite direction. Enzymes can be bound to nonreactive materials, like small globules of plastic or ceramic or a fine mesh of some kind. In this immobilised state they still function as catalysts and they don't get destroyed in the process of production. One can even immobilise enzymes within cells, which survive as barely functioning, inert capsules, locked in useful immobility. Proponents of the use of immobilised enzymes also point to the fact that they can be used as chemical sensors, registering changes in acidity, temperature and other parameters, and signalling these changes to a control system. One could, for example, have a membrane that modified its own properties in response to the surrounding conditions.

The markets for detergents, meat tenderisers, products for jam- and cheese-making and brewing industries and sweeteners are not small. But a route into a far larger arena is suggested by an idea from the Cetus Corporation for making basic ingredients from sugar for the plastics industry using biological enzymes. Essentially, this is a process for making a substance called propylene by dovetailing two sets of reactions, one that ends with the oxide of propylene, the other which converts the sugar, glucose, to its more useful relative, fructose. This is technically very neat, and it is the use of bacterial enzymes that makes it possible. The elegance of the idea is enhanced by the fact that it also yields glucosone, which can be used in washing-up liquids, and fructose, which is the sweetener we have already met. In principle, then, this could be one way into the plastics market which is worth $50 billion a year. In 1981 a Cetus spokesperson estimated that $2 to $3 billion worth of propylene oxide might be made this way by the end of the decade. One can't help but notice, however, that in mid 1982 Standard Oil of California (Socal), who were putting up the money for the development work on the glucose-to-fructose process at Cetus, with which this idea is surely associated, pulled out of the arrangement. So perhaps things are not that straightforward technically. But whatever was going on there, it is a graphic example of how one can get from new feedstocks to plastics with biotechnology. In this case one has to start with propylene, and large quantities of glucose, probably from starch.

Let us now look at another set of processes that yield what are known as aliphatic organic compounds, which include solvents like ethanol and organic acids like acetic acid. Many of these compounds used to be made by fermentation. Ethanol or ethyl alcohol is grain alcohol, and any home wine-maker knows that yeast will produce vinegar if you let it. Ethanol can be used as an industrial solvent, rather than as a potable alcohol.

Actually it has an enormous range of uses, in antifreeze, and in making other solvents, extractants, dyes, drugs, lubricants, adhesives, detergents, pesticides, plasticisers, surface-coatings, cosmetics, explosives and resins for the manufacture of artificial fibres. No wonder then that in the US, 619,000 tons were produced in 1980, and brought in $297 million. Fermentation ethanol can be produced from sugar-beet or cane molasses or from starch, obtained from maize, wheat, rye or cassava, or from cellulose. It is said that the prices of starch and sugar fluctuate too widely for a fermentation industry to be established upon them. Be that as it may, large programmes are being set up in some developing countries with the specific aim of producing ethanol from sugar-cane and cassava. The best known example is in Brazil, although there are others in Indonesia, Kenya and elsewhere. The fact that cassava is a staple food has not deterred investors. Rather it has spurred research into improving the resistance of cassava to disease. The International Plant Research Institute in San Francisco, in collaboration with the process engineering firm Davy McKee, is now developing new cassava strains specifically for commercial fermentation. Whether there will be any spin-off from this plant breeding for millions of peasant farmers who grow cassava as a food of last resort remains to be seen.

In general one can say that although land is often needed to produce food for local populations, the control over land use is such that it is often used for cash crops like coffee and rubber. Rural communities, which are frequently displaced from the land by large estates and plantations, derive little benefit from them. So it is likely that the pattern will be repeated with "fuel" crops and "fermentation" crops. If there is sufficient land to produce the necessary food, then turning some of the surplus ground over to the production of crops for fermentation may be a good idea, saving valuable foreign exchange and building up indigenous scientific and technological expertise. If organic material that would otherwise be wasted is used, then there would be obvious gains and fewer possible drawbacks. There are, however, some real problems concerning the politics of land and energy use. We come back to this later.

Let us assume that for these kinds of reasons wood is a preferable starting material. The problem is to break it down so that the cellulose is available in a pure form. This can then be broken down into glucose, since cellulose is a polymer of glucose which can then be turned into alcohol. This is a two-step process. An improvement would be a one-step process, in which pretreated wood is fermented directly to alcohol. There are some marvellous micro-organisms which can do this although they are rather rare and not very well understood.

An alternative, that ought to be predictable by now, is to attempt to create a yeast or a bacterium to order, by splicing new genes into it, which can ferment cellulose directly to alcohol. Also one can concentrate on another wood component, xylan, which is a polymer of the sugar xylose. It turns out that if xylose is converted to another sugar, xylulose, it can be fermented to ethanol by yeast, so one idea is to get the genes for xylose isomerase, the enzyme that converts xylose into something that yeast can use, into yeasts. Or you can simply keep on looking for new bacteria, since there are thousands of unknown strains with all kinds of tricks all around us.

For example, for thousands of years yeasts have been used to make alcohol. But other organisms will also do it. The micro-organism *Zymomonas mobilis* is used to make the Mexican drink pulque, and it is twice as efficient as yeast. There is no doubt that in the longer term cellulose will be processed into chemical feedstocks such as ethanol and more complex materials such as polymer fibres and films. Already Gulf Oil Chemicals and the Raphael Katzen Association are building a plant that will make 150,000 gallons of ethanol a day from cellulose. The capital cost is $112 million, which indicates a major commitment to the process. It is worth bearing in mind, however, that while ethanol production is one of the more energy-efficient ways of dealing with wood waste, perhaps only a quarter of the residues could be collected. The net energy gain from producing ethanol as a fuel in the

US, when the energy expended in collection, processing and heating is taken into account, would amount to only about one per cent of the current US gasoline consumption. Its use as a chemical feedstock is another matter.

Organic acids also make up a sizeable chunk of chemical production. Acetic acid has all kinds of uses, apart from transforming the taste of fish and chips. It is used in making rubber, plastics, acetate fibres, pharmaceuticals, insecticides and photographic materials. Excluding the amount used as vinegar, 1.4 million tons are made in the US every year, bringing in $500 million. Some work is going to make acetic acid by the fermentation of cellulose, which is thought to be a cheap feedstock that could compete with oil; or it may prove possible to get bacteria to synthesise it from hydrogen and carbon dioxide. Another acid that is used in vast amounts is citric acid. The world market is 175,000 tons at a value of $259 million. It is made by fermenting molasses with the bacterium *Aspergillus niger*. Cellulose might be a cheaper substrate, and work is going on reprogramming *Aspergillus* with the genes that specify cellulose-degrading enzymes. Lactic acid can also be made by fermenting sugar, and about half the European supply is made this way, although it turns out to be very expensive to separate the acid from the bacterial culture in which it was produced. Lactic acid can be made into lactide, which in turn can form long chains, very like those in better known polymers, polystyrene and polyvinyl chloride (PVC).

One could go on at some length. The list is not endless, but the basic principles are the same. The current theme is the substitution of a new feedstock for oil, a new, simple hydrocarbon from a living source to take the place of one entombed in fossil reservoirs for millions of years, even though now oil is still cheap enough. Another theme is the idea of simple chemical transformations that add and subtract building blocks or allow structures to fold up or form chains. Micro-organisms can process hydrocarbons without much sweat, although they sometimes poison themselves with the products of their labours. That is the essence of microbial biotechnology.

The last set of fermentation products we shall consider are the amino acids. We have already met these as the basic constituents of protein molecules. But even elementary building blocks have to come from somewhere. Organisms may make some of them themselves from simple nutrients, or they may have to be supplied in the diet, in the form of protein. In the case of human beings we need to obtain eight amino acids in our food. Consequently they tend to be added to food that would otherwise be deficient in the needed amino acids. Lysine and methionine are sold as animal feedstuffs additives; glutamic acid is made for the production of the flavour enhancer monosodium glutamate, and fermentation is used to make thousands of tons of these substances per year. Again genetic manipulation is being used to trick them into making more.

Money Does Not Grow on Trees, But in Them

There is a nice story that appeared in the *Financial Times* (and so must be true) of a senior research scientist at Exxon saying in the 1960s that oil was a far too subtle and interesting chemical cocktail for it merely to be burnt. For that chemical reverence he got into trouble because, after all, much of the turnover of Exxon, larger itself than the GNP of many countries, comes from selling oil to burn. Releasing the energy locked up millions of years ago in plants by combustion is one of the cheaper ways of producing it, but as everyone knows, particularly the oil companies, global oil reserves are finite. With the prospect of serious shortages in view, the oil corporations are diversifying into other forms of energy, buying mining companies and investing in the nuclear industry.

However, having got organised around the immensely profitable global transport, distribution and refining of a hybrid energy source, there is a certain desire to stay with that kind of product, not least because other technologies like the motorcar are dependent on it. Personal transportation could be re-engineered around, say, a fuel cell or a battery charged up from the grid, but a liquid, combustible, hydrocarbon fuel requires less of a change. The neat thing would be to use renewable sources, which can be steadily burnt and replaced.

This pressure on energy prices has intensified research on the utilisation of what is called "biomass" as an energy source. "Biomass" is energy-rich material that is biological in origin, such as felled timber, forestry wastes, harvested sugar cane, wheat straw, bracken, plant seed oil, seed husks, the residues from the processing of plant materials in a paper factory, algae, sewage and animal slurry. There is a vast range to choose from and much of it is simply thrown away at the moment. Research on getting energy from biomass is not new, and most of the ways of trying to tap its energy content are largely established, if not ancient and traditional like fermentation. What is new is the intensity with which these forms of biotechnology are being pushed, and the way in which sophisticated biological and genetic techniques are being brought into the development process. The tendency all the time is to engineer new levels of capability into business materials by getting at their genes. I am going to consider four different forms of biomass energy production: using hydrocarbon latexes as a liquid fuel, fermenting organic material to produce a fuel, using oil seeds as a fuel source and biogas produced by anaerobic digestion.

The first of these is the most startling. The rest of them are perhaps less surprising. Anyone who has put brandy on a Christmas pudding or raced model aeroplanes knows that alcohol will burn, and anyone who has stirred up an old pond knows there is gas at the bottom from rotting vegetable matter.

Many plants produce latex, a sticky fluid that can be tapped from the trunk. The example of this which is best known to many people is natural rubber, which is made from the latex secreted by the rubber tree, an emulsion of long, springy molecules in water. Many other trees in the same family will do the same thing, and one of them, *Euphorbia latyris,* has recently begun to attract a lot of attention. This tree has its champion, a Californian scientist called Melvin Calvin who already has a Nobel prize for his work on photosynthesis (converting the sun's energy into plant energy). *Euphorbia latyris* is a bright green shrub, that grows to about six feet in height and is not particularly demanding about the soil in which it grows. In America it is known as the gopher plant, since its root system is thought to stop gophers from criss-crossing one's land with underground tunnels. A convenient way to extract the latex is by harvesting the plant and processing the dried material with acetone or benzene, ending up with a liquid that is not unlike petroleum in its mixture of hydrocarbons. Calvin decided to test its potential as the basis of an "energy plantation," and his first results suggested that one could obtain around twenty-five barrels of latex per year from one hectare of plants, at a cost per barrel that he put at $25. That compares favourably with crude oil. But cost estimates of this kind are notoriously shaky, and commercial production of *Euphorbia* "oil" would, it turns out, require vast tracts of land.

Ernest Bungay analyses the problem as follows. One can assume that an acre of land would produce ten tons of dry biomass, of which perhaps ten per cent would be "oil." One ton of oil equals about seven barrels. If cultivation costs $150 per acre, the unprocessed oil would work out at around $20 a barrel. The yields obtained from commercial plantations would, in fact, be much less than this since the plants are much more carefully treated in experimental plots and laboratory extraction is likely to be very thorough. It has also been suggested that where the land is available the rainfall is too little. Calvin has argued persuasively that yields from rubber trees have been boosted by twenty times over the last fifty years, with a wide range of *Euphorbia* varieties growing in different parts of the world. There is good reason to suppose that significant genetic improvement in yield, oil composition and resistance to disease could be obtained. The techniques discussed in the last chapter suggest how that process might be speeded up. I don't know if anyone has yet cloned *Euphorbia latyris* in tissue culture. Also, the biomass remaining after the latex has been extracted could be turned into ethanol by fermentation.

Let us now consider the land requirements. To produce 1000 tons of biomass per day, with an output from the processing plant of 700 barrels of oil, would require an area of some 300 square kilometres, which is roughly the area enclosed by a circle of 10 kilometre radius. You could fit a city the size of Cleveland into that fairly easily. But *Euphorbia* will grow successfully only in hot climates, and in Arizona, New Mexico and Nevada there is a great deal of uncultivated land on which the shrubs could be grown. The world's most productive weapons laboratories and missile test sites are successfully lost in the vastness of their deserts and mountains. One estimate suggests that a processing plant delivering 50,000 barrels a day, a minimum import level for many refineries, would require 4000 square kilometres of land. That is with a greater productivity than we assumed above. Apparently hundreds of thousands of square kilometres are "available" in the southwestern United States, but to buy that amount of land or the right to cultivate it would presumably necessitate a massive capital investment. Nonetheless, oil companies have begun to buy enormous tracts of land for extracting oil from shale. One then has to consider how you would harvest gopher plants at the required rate. This kind of calculation suggests that the gopher plant is not yet a viable proposition as an energy crop.

But there are other possibilities. The Brazilian tree *Cobaifera langsdorfi* grows wild and produces twelve gallons of oil per year when it has reached maturity. The tree can be tapped, much as sugar maples are, to give ten to twenty litres of oil in a couple of hours. The liquid is sufficiently close in composition to diesel fuel for it to be burnt in a diesel engine without any further processing. Before very long, however, the engine gets clogged by gum, but refining the oil might remove this problem. Then there is *Croton sonderianus,* the black quince, which grows as a weed in northeastern Brazil. This member of the *Euphorbia* family is thought able to produce at least fifteen tons of dry plant material per year, and perhaps three times that. If the biomass is treated with steam, a useful hydrocarbon fuel can be distilled off. Another possibility is eucalyptus oil, which can be used to run an engine on its own or in a 70/30 oil/gasoline mixture. One estimate in 1979 put the cost of the oil at $35 per gallon, which is far above what anyone pays for gasoline, but there may be scope for improvement here. Yet another plant with a future possibility is the common milkweed *(Asclepsia speciosa),* which is widely distributed around the world. Test rigs in Texas and Alabama suggest that mechanical crushing and refining would give a fuel at a price competitive with gasoline.

These hydrocarbon-rich liquids don't have to be burnt. The latex from guayule *(Parthenium argentatum)* can be turned into rubber. Indeed the use of the plant was brought forward when the Japanese seized the rubber plantations in the Far East in the Second World War. Goodyear has made radial tyres from guayule rubber that pass the American Department of Transportation high-speed endurance tests. Similarly the jojoba *(Simmondsia chinensis)* is exciting interest as a source of oil that can substitute for sperm whale oil. Several thousand acres are already under cultivation in Arizona and California. Preliminary analysis suggests that commercial cropping, with thorough seed collection to maximise the yield of oil, would have to be done with care, since even deserts have ecosystems which could be hammered by intensive harvesting.

So far we have been talking mostly about hydrocarbons extracted from the whole plant. In seeds, the oil content is often higher and it is easily extracted by crushing. In the Second World War the Chinese developed a process of cracking vegetable oils to separate out the more volatile fractions. They used oil from tung nuts, rapeseed and peanuts. On the other side in that war the Japanese used the seeds of the "petroleum nut tree" as a fuel for their tanks. Interest has recently swung back to oils like these. Soya bean and sunflower seed oil have been tried out, with yields of at least one tonne per hectare. The cost works out at around $2 a gallon and the net energy gain of available fuel energy per unit of energy expended in harvesting and processing is between 3:1 and 10:1. That is an acceptable gain, but often the problem with schemes of this kind is that you end up with less energy than you put in.

In the last chapter we considered the use of palm oil to make margarine and detergents. But it can be used as a fuel, igniting under compression as diesel fuel does. In Brazil, a country with a massive programme of research into alternative fuels, a whole series of new fuel mixtures is being tried out, using commercial and public service vehicles. The National Institute of Technology in Rio de Janeiro completed trials in January 1981 of 4/1 diesel/peanut-oil mixture and a 73/20/7 diesel/palm oil/ethanol mixture. Interestingly, in over 10,-000 kilometres one bus had a fuel consumption that was 3.4 per cent better than on its normal diesel fuel. The plan is that sixteen per cent of the demand for diesel fuel in Brazil will come from vegetable oils by 1985. Their use of fermented alcohols is considered below.

In the consideration of energy plantations above, the assumption was that the plants would be harvested commercially and the oil sold on the open market. This led to the idea of vast plantations, of unknown economic viability, in order to service large refineries. But oil crops could be used by individual farmers to attain self-sufficiency in fuel. In South Africa intensive research is going on with sunflowers, and yields are already up by 2½ tonnes per acre since 1970, with 500,000 acres in cultivation, three times more than a decade ago. New hybrid seeds have produced yields of four tonnes per acre. In the US, similar yields have proved feasible, with two or even three crops per year. Oil yields work out at 100 gallons per acre so putting, say, twenty acres of marginal land down to sunflowers and investing in an oil press would yield a substantial amount of fuel. I am, of course, assuming here that farmers will have twenty acres to spare. That is a small amount for a farmer in the developed world; it is a vast amount to millions of subsistence farmers, who do not use tractors anyway.

Plant oils represent a class of alternatives to fossil fuels. Alcohols are another, and one that is more appealing at the present. When discussing feedstocks, we considered how wood cellulose or sugar-cane juice could be fermented to ethanol. For a long time it has been known that ethanol will also serve as a very acceptable fuel. Henry Ford I was much taken with the idea that it would supplant fossil petrol in time. A major fuel plant was built in Kansas in 1936.

The most striking development of this kind in recent times is the Brazilian alcohol programme, begun in 1973, which is mostly devoted to producing ethanol fuel supplements for gasoline, though ethanol is also to be the feedstock for the Brazilian chemical industry. The US government has also established a gasohol programme, working with corn, which has not been very successful economically. The Brazilians on the other hand are prepared to subsidise their fermentation programme because of the massive potential savings of foreign exchange.

Brazil's foreign debt is now some $64.5 billion, which makes it one of the larger in the Third World. It costs $19 billion each year to service a loan of that size. The balance of payments deficit was $12 billion at the end of 1980. In that year the cumulative expense of importing 670,000 barrels of oil per day added up to forty-six per cent of the value of all imported goods. Clearly, then, a situation like this creates enormous pressures to find a substitute for oil, and the national plan is to limit oil demand to 1,500,000 barrels per day by 1985, of which only 500,000 will be imported. Of the fuel produced domestically, 500,000 barrels per day will be oil. Sugar-cane and wood alcohol are projected to supply the equivalent of 170,000 and 120,000 barrels per day or twenty per cent of the total national energy demand. One estimate suggests that to replace one million barrels of oil per day would require twenty synthetic fuel plants at a cost of between $1 and $2 billion at 1979 prices.

Various crops can be used for fermentation, including sugar-cane, sorghum, cassava and maize. In Brazil 100 million tonnes of sugar-cane are grown each year on 2.5 million hectares. Twenty per cent of the crop goes to alcohol, most of the remainder is processed into sugar to be sold on the world market. Cassava is also grown on two million hectares of lesser quality land. This produces thirty million tonnes, the largest cassava crop in the world, most of which is used for food. The production of alcohol from cassava in Brazil is not great at the moment, though it is likely to increase substantially. Some people see it as an appealing alternative to sugar-cane, since it can be grown on much poorer soils. It provides four times more job opportunities, because it needs annual replanting, and it is much less dependent on the world market price.

The Brazilian national alcohol programme is extremely ambitious, and some commentators are sceptical about its present performance and the attainment of the projected levels of production. One aspect of the plan is that by 1985, 2,121,000 vehicles are to have been converted to run on 100 per cent ethanol. One problem is the fluctuating price of sugar. In the late 1970s sugar prices rose dramatically. The savings on imported oil from the alcohol programme in 1979 were $300 million, but if the sugar-cane involved had been processed to sugar it could have been sold for $1.5 billion. That kind of disparity does not seem to encourage government investment in distilleries rather than sugar refineries. Another problem is land use. If alcohol production is to increase threefold by 1985, a prodigious amount of additional land will be required. This is likely to lead to the expansion of large plantations owned by the sugar companies at the expense of smaller companies. Similarly if cane yields are to be increased, a significant increase in the use of fertiliser (which has to be imported) is likely. One ICI commentator has noted that plant hormones could speed up the growth of sugar-cane, which is already the most efficient photosynthetic converter known. Hormones like the gibberellins used to be expensive, but with reprogrammed bacteria they might be made more cheaply. But distillation of alcohol also produces a great deal of waste. Each litre of alcohol comes with twelve to thirteen litres of acid liquor that is often discharged into local waterways with disastrous results.

The alcohol programme may be a remarkable experiment, in which the technocratic government of a Third World country successfully manages the problems of rapidly rising energy demand, import substitution and rural unemployment created by cash-crop agriculture. It may be a route away from a crushing dependence on imported fuels, using the vast resources of land and sunlight that a country like Brazil possesses. Only a vast, sparsely populated country can even contemplate a fuel programme of this kind. If you assume that a hectare of land will produce, say, twelve tons of biomass per year, then you can calculate what proportion of the national land area is needed to grow the national energy requirement. I am rather struck by the exact figure of 100 per cent for Italy, which conveys an image of the whole of Italy neatly planted with sugar cane right up to the coastline without any room for motorways, cities, factories—or people.

On the other hand the alcohol programme in Brazil can also be seen as a subsidy to an industry that is not very efficient or innovative, which has an appalling pollution record and which will push the trend to monoculture further in the countryside, to fuel the cities at the expense of peasants, some of whose land will be taken from them. Commentators have a habit of saying that land used for food must not be turned over to fuel, but the clear lesson of history, from the enclosures of the fourteenth century to rubber plantations in the twentieth century, is that land use is determined by land owners seeking to maximise the return on their investment property, in defiance of those whom it used to feed.

The Effluent Society

Farming produces waste; food processing produces waste; fermentation produces waste; human beings produce household waste and sewage; many industrial processes produce effluents. An exciting possibility from biotechnology is that these streams or piles of material could be processed into food, alcohol or useful chemicals. For example forty per cent of human excrement is usable protein. In the Western developed countries we simply throw that away all the time, even though processes exist for recovering it. In less wasteful countries, with different sensibilities, it gets used as a fertiliser—it's called "night soil."

Organic effluents are food for some organisms. Much urban sewage is broken down by bacteria and algae which are encouraged to live in settling tanks through which the waste material flows. Most effluent in the UK is treated by the activated sludge process, first used in that centre of inventive activity, Manchester, in 1914. Treatment systems like this merely transform the effluent into a form that can safely be discharged into waterways or on to the land. But one can also think about using waste as a growth medium, or as a way of producing gases like methane, or for fermentation or combustion. Various processes now exist to transform household waste into an oily fuel that can be burnt in a power station. Plant waste, like the "bagasse" left over from sugar-cane harvesting, can be burnt efficiently as a complementary energy source at the distillery. Papermaking liquor will support bacterial cultures to make single-cell protein, and farm wastes can be digested anaerobically by bacteria that give off methane. One estimate puts the number of such gas installations in China at seven million. In Europe, they are not yet appealing since the net gain of energy over the year is not that great, given the capital cost involved; but an increase in the costs of pumping slurry into the sewage system could change this situation.

Not only do frozen pea, jam and instant mashed potato factories put a lot of sugar and carbohydrate down the drain, but a great deal of metal also finds its way into the sewers, e.g., zinc, copper, cadmium and mercury. Some of this can be recycled by biological processing. For instance, there exist bacteria that will concentrate copper. If they are grown in a culture containing copper salts they will gradually build prodigious amounts of it into themselves, removing it from the surrounding medium. Similarly, one can pass industrial effluent such as the acetate residues from rayon-making through a fibre matrix and concentrate zinc. There are even bacteria with a passion for uranium, which they will slowly extract from sea water. The Indian microbiologist Chakrabarty, once employed by General Electric and now immortalised in patent textbooks, has created bacteria that will happily live on effluents from chemical plants and others that will break down the herbicide 2, 4, 5-T, such is their relish to deal with unpalatable materials.

Waste processing of this kind clearly has a future, even though many of the ideas are uneconomic at present. The problem though is to ensure that it is not used as a technical fix to clean up wasteful, polluting, energy-intensive industrial and agricultural processes, but that it is built into the design of the mode of production. It could be placed at the heart of how our society processes "renewable chemicals."

So?

What does all this mean? First, the massive stresses now building up in an industry with a global reach and a history of confident growth could produce some interesting developments. Something must happen to bring down the costs and consumption of energy, and that will mean operating chemical plants at lower temperatures and pressures in some cases. Similarly, if new feedstocks can be produced by biotechnology, then the industry can begin to recycle waste materials or to work with renewable chemicals rather than burn fossil reserves. That has a certain appeal on the face of it.

Second, it is clear that there is no single cheap and easy solution. All the technical options that we have considered could only be implemented at the moment from the margins of the industry. Their technical or commercial uncertainty is too great for an immediate, massive commitment of resources to be made. In the longer term new feedstocks and processes must be found, but the gap between future possibility and present reality is very great. It can be closed only with a great deal of money.

Third, it follows from this that whatever options chemical companies select are likely to be pursued with considerable vigour, even if they create major social and political problems. The fact that it takes a great deal of land to grow enough plant material to service a chemical plant does not mean that companies will hold back, if and when they feel it will serve their purposes. One can confidently predict that land use—for fuel, feedstock and food—will become a major political and economic issue in the next two decades. Gasohol programmes in developing countries are already starting to change the orientation and structure there. Some of that alcohol will feed local chemical companies, based close to the new feedstock plants. But of these some will be wholly or partly owned by the existing foreign multinationals. The twenty-first century's analogy to the land enclosures of earlier periods may be to keep the consumers of the developed world supplied with washing-up bowls and cling-film.

Similarly, it is likely that the politics and economics of waste and recycling will increase in importance. The diversity of biotechnological processes means that sugar cane or corn cobs or forestry debris or weeds can be turned into a variety of materials, such as food, biological fertiliser or alcohol. The question of which usage is promoted should be regarded as a fundamental political issue, since it concerns the appropriation of a valuable social resource. After all it is likely that it will be taxpayers' money that will bring the relevant biotechnologies on stream.

All this means that the kind of chemical industry that will exist in the twenty-first century is under consideration now, but in circumstances that are much too exclusive. Its future is of immense importance economically and politically, not only to shareholders and those whose pensions depend on its profits, but also to workers in the industry and to consumers. It is hard to think of an aspect of contemporary life that does not depend on its products. Something as intricately woven in the very fabric of social existence ought to be given much more public scrutiny and appraisal. Yet the opposite is the case at the moment.

Multinational chemical and energy corporations are the epitome of closed, secretive institutions, whose very idiom of decision-making is unknown to the public. They have the resources to simulate and commission technological futures, thinking a decade or two ahead without consulting those whose lives they will reorder. The industry itself supplies us with the image of a pipeline that captures the idea of new products and processes flowing out from some remote decision centre. When these meet the light of day—when the new plant is put up or the new product marketed—it has built into it forms of work organisation and definitions of need, which it is then far too late to challenge.

Through the private constitution of technologies, corporations steal time from the public—time that they use for research and development, for appraising options, for deciding where, with what materials and with what kind of workers new products will be made—time that could be used to debate these choices more widely within the companies and more generally with local communities, nationally and internationally. That silent, selective appropriation of the future could not happen without the eager assistance of an army of mental labourers, the scientists and technologists who feed ideas into the pipeline. Getting them to rethink the terms on which they are designing a future is therefore an important theme in my final chapter.

7

Where Do We Go from Here?

Smuggling the Future Out of the Boardroom

Over the last six chapters I've been trying to evoke the power and implications of biotechnology. I've been trying to sketch a technological and industrial image of power, linked to conceptual and perceptual changes of tremendous subtlety and force. I want to be able to say, "Now, don't you see? It's like that." For I think that we are living in the early phases of a major industrial transformation, a revolution perhaps. I think that our attitude to and interaction with nature are being recast, and I do think that the energy that is behind this process is awesome. The problem is that the phenomenon under analysis is exceedingly complex. It is also emergent and fluid, its nature and direction are not yet fully manifest.

At one level I find this process of industrial restructuring and the signs that something new is appearing endlessly fascinating. It is a regrouping of forces that doesn't happen that often. We have a rare historical opportunity to watch it all in motion, to see the technical skills of gene-splicing being used to unlock another cycle of industrial change. It is obvious that even if the first investment bubble has burst —and we'll return to that point later—the technical advances are here to stay and that they will continue to alter decisively our notions of what is possible scientifically and industrially.

Once you have begun to leap over the limitations of particular species, by programming in something that another one can do, that lesson is not forgotten. As a result our attitudes to what bacteria, moulds, yeasts, plant cells, plants, mammalian cells, farm animals, human cells and human beings can do are being shifted. Functions are being disembodied or decontextualised from a particular species, so that they can be relocated where it is most useful or profitable to express them. Species and function now have no essential connection. Their association is contingent, dependent upon the will and ingenuity of the genetic engineer. Species are now those divisions in the living world that we choose to accept. Far from nature appearing, as it did in the eighteenth century, as a grand canvas of organic forms, a hierarchical Chain of Being filled with the products of divine artifice, we are learning to understand it as stockpile of programmed systems, offering a comprehensive range of components and modules to the corporate customiser.

A realm of planned combinations has been revealed as an ambiguous part of "nature," being simultaneously natural and artificial. Within it organisms are at once "products of manufacture," "man-made," hybridised to an exact specification and owned as private property, and yet also part of the panoply of living things. A new dimension of possibility has appeared.

So what kind of metaphor do you use to capture the texture and significance of this process of technological change? Clearly it is something like a revolution. It is a major economic phenomenon that will have social and political repercussions. It will affect the patterns of trade. It will change the value of some people's assets, both positively and negatively. It will force some industries to the wall. It will have profound effects on the global structures of power. That is why intelligence agencies like the CIA send their representatives along to business briefings. After all, if you can grow opiates in cell culture, what will that do to the heroin trade in Southeast Asia or Turkey or to the cocaine business in Colombia? Or if single-cell protein does turn out to be a cheap animal feedstuff, will the Soviet Union need to buy so much grain from overseas? Or if corn syrup replaces sugar as a sweetener what will happen to those Caribbean islands where profit from sugar is fundamental to the suppression of dissent?

Yet for all this, it is not a revolution in the usual political sense. It is not an uprising, guided by a vanguard party, or a process of political liberation that bursts its intended bounds of reform. It may cause governments to fall, but not directly through the loss of political legitimacy. It will be because innovation has set loose economic forces that they are unable to contain. Actually, I think it is more likely to keep precarious élites in power, by offering them new links with oil, chemical, food and pharmaceutical companies, looking for land on which to grow chemicals or wastes to transform, or markets for new products like hepatitis vaccine. In this context biotechnology will appear as a technical fix for some economic problems. This will appeal to the professional middle class, and benefit the owners of local industry. That is how one might see the gasohol programme in Brazil, which has opened up opportunities for foreign motor companies like Volkswagen, chemical companies like Dow and Mitsubishi. Gasohol could help the millions of motorists in Sao Paolo and provide another line of argument for further rescheduling of the continually growing mortgage on the country's assets.

One of the problems, of course, is that our everyday experience and language do not contain the forms that can match the complexity and global reach of what's happening. We don't have the idioms with which to represent to ourselves the process of change. We reach for terms like "revolution," "explosion" or "a new frontier," in the hope that they might allude effectively to analogous events. But they are threadbare terms, not least because the planning and management of such periods of innovation is such a remote activity.

The chairman of ICI, John Harvey Jones, recently locked himself away in the Millbank Tower to produce a new plan for the corporation. When he emerged, he had decided to sell the company headquarters and to expand activities in the US market with promising new specialty chemicals pulled from the company's streams by a special new team of "fishermen." Multinational corporations remake the world economy in very cloistered circumstances, giving out signals as they do so that competitors and investment analysts can read, consulting governments over subsidies and financial policy, and occasionally buying in a little advice or some new ideas if their internal sources need corroboration or stimulus. As far as most of us are concerned, it could occur on Mars, until the decisions are announced and the effects of "rationalisation" become clear.

But this is how biotechnology is being created, behind closed doors. The result is that we simply don't know how to conceptualise what is happening. It is too ramified, too interconnected, too remote from our experience. Most of us simply don't know how governments, large corporations, financial institutions, venture capital investors and scientific entrepreneurs interact. Indeed I suspect that it is only when major technological issues like this reach the mass media, that most people even come to hear of such things, let alone learn how they function to bring specific technologies into existence. But while we marvel or shudder at the sensational reports of *Tomorrow's World*, people are working late at the lab, in the office or at the bank to bring on stream projects that will only surface in the public domain in five, ten or twenty years' time.

If that seems an exaggeration, remember that Unilever's research group began work on oil palms in 1968. Some of the projects at G. D. Searle, the American drug company with a biotechnology lab and production plant at High Wycombe, go back to the late 1960s. ICI's Pruteen plant goes back to the mid 1960s and, because of the swings and roundabouts of the international feedstuffs and oilseeds markets, may only pay for itself as an investment in fermentation expertise and sophisticated microbiology for the chemical industry of the 1990s. Plans are being laid now that will reach into the twenty-first century and wipe out the divisions between whole industries, such as those between energy and agriculture. A new global web of economic and technological relations is being spun, using the first fruits of patronage, via the products of contract research of the genetic engineering companies.

That process is how, in industrial society, technological change is managed, and how the thrust into the future is likely to reproduce gross inequalities of wealth, status and power. The future must be smuggled out of the boardroom and covertly put into practice, otherwise it would be challenged by those who continue to lose out and are short-changed by the mode of production. The appalling thing is that many of the people involved in this exercise would be more than happy if things were carried on like this with planning and decision-making coming first, and consultation, if necessary, afterwards. After all, scientists were not alone in pressing for the exercise in participation concerning recombinant DNA to be wound up. "Enough," industrialists were also heard to say, "is enough." Commercial secrecy requires that certain issues are not aired in public, they argued, just as their representatives have argued against EEC legislation requiring multinational firms to disclose commercial data to workers. But if we accept that, investment, product choice, location of manufacturing plant, and diversification are not even negotiable. They are effectively unknown, except to remote members of corporate boards, until they appear as *faits accomplis.* The terms in which the future is conceived, the priorities, the options, the areas of uncertainty and controversy, and the value placed on skill, employment and autonomy, remain a mystery, as do the identities, competence and grasp of the people taking the decisions.

One danger with this kind of account is that one tends to lump together different companies, industries and sectors of the economy and grant all their executives a depth of perception and resolution that they do not have. One tends to see "strategy" everywhere. Making one hermit, shunning the world in the ICI Millbank Tower, stand for a company is all very well, but corporate development and investment cannot be that deliberate, synchronised or efficient. Nor can the path into the future ever be that plain to see. It is more a question of balancing uncertainties and reaching a consensus about what might work for a particular company. Nonetheless, the point still holds that biotechnology is being developed through thinking in private. Similarly it is all too easy to fantasise about the well-calculated movement of millions of dollars, yen, deutschmarks and pounds into new projects, according to meticulously thought out schemes from financiers and investment managers. In fact the frenzy to buy shares in Genentech and Cetus and similar, though less spectacular, flotations in Britain suggests that *some* investment is anything but rational. There has been a Gadarene rush to own stock in an industry that has very effectively sold an image of itself as glamorous, fast-moving, inventive and incredibly profitable in the medium term.

Not all investment in biotechnology is like that. In chapter 2, I mentioned the stages through which these small companies pass. With each there must be some essentially private bidding and appraisal. The deal is sometimes not struck, as enthusiasm is switched off. Investors do pull out and companies do fold, or fire a third of their staff—or, as in the case of DNA Science, the cluster of companies that was to be set up by the Wall Street firm of E. F. Hutton, companies are abandoned when corporate sponsors want too great a return. The whole exercise can be restructured as a tax shelter. This is to say, even when company formation goes wrong, nimble financiers can reconstitute the assembled expertise to run an investment machine that steers spare capital safely away from taxation.

In the UK, the Prudential Assurance Company has set up what is in effect a venture capital operation, called Prutec, which is cautiously and quietly sifting through, selecting and investing in high-tech projects, not only in biotechnology. Nothing Gaderene about that. That seems to be the style of the British Technology Group as well. (The BTG incorporates what used to be called the National Enterprise Board, set up in the mid 1970s as a kind of state-owned merchant bank to gain a degree of political control over capital investment in key areas of British industry. Its role now seems to be far more concerned with the facilitation of private investment.)

We can view this governmental activity in at least two ways. It could be a way of underwriting industrial investment decisions, using taxpayers' money to stiffen the resolve of large private investors, to their probable mutual profit, until the public holding is sold off. Or one can see it as a way of getting industrial validation of decisions that concern public money. Governments do, after all, have a way of wasting money on high technology. *Either way* it means that nothing happens unless the project under scrutiny fits very conventional criteria, acceptable to corporate investors, about what constitutes the right kind of investment. No one gets access to capital unless they are wearing the right kind of lab coat, at least if you go to the usual sources.

Innovation as a Race

The focal point of what I have been saying so far is that industrial dreams for biotechnology and their interpretations by bankers, when confronted with a financial reality-principle, are a very private business. The dialogue is *sotto voce,* discreet and effective. The money appears, the teams are assembled, the fermenters built, and the marketing planned. Often only at this stage is there any hint to the public that a new process will shortly arrive. Indeed by this stage the priorities have been settled for some years, the goals already set up when the exploratory research was done. Rather, it is the beginning of the process that we have to focus our attention on if we want to change the ways in which, and the purposes for which, biotechnology, or any technology, is created. If we leave that phase alone, exempting it from scrutiny and analysis because of the belief that the practice of research is somehow separate from its application, then we lose the chance to intervene later on, in the process of development and commercialisation. That, in a way, was why the moratorium on recombinant DNA research was such an interesting and potentially valuable experiment in public participation. That is why it is so sad that the debate was shunted on to the sideline of potential hazards, whilst undebated activity of planning and investment went ahead on the main line. That is why it is worth concentrating on the formation of biotechnology by laboratory scientists, who use public funds. It is this community that is busy developing a close relationship with private industry, at such a pace that it will soon be impossible to pry them apart or to suggest that their cohabitation is anything other than natural, healthy and mutually satisfying.

I would like to think that biotechnologists could have other partners, and that they would get recognition, respect and new ideas from a change of orientation. At the moment, biotechnologists in Britain seem to be trying hard to catch the eye of government, to obtain the resources for expansion, only to be met with polite interest but no real generosity. There is no doubt that British biotechnologists feel more than slightly beleaguered, underfunded, misunderstood and envious of those colleagues who have already taken off for America or Switzerland. How long will it be before there are stories of leading researchers decamping for Saudi Arabia or Brazil or South Korea? One of them said recently that "the situation has gone from abysmal to bad." That was as cheerful as he could be. All this, because biotechnology is a field where large amounts of development money are essential to turn laboratory ideas into commercial processes, over a time scale that deters all but the most resolute and imaginative private investors. Moreover, British biotechnologists look to other countries, like Japan, West Germany, France or the USA, knowing themselves to be national representatives in a technological race, and point in despair at the scale of support that their competitors receive.

Whether or not they all feel equally concerned about this I don't know. Certainly some do. I want to let their attitudes stand for a particular view of technological innovation as a race between research groups in which there are winners, who get recognition and funding all the way to the marketplace, and losers, who wander through the corridors of power trying to find someone who will listen to them. The race may be not simply between particular labs or companies but is more likely to occur at the national level, with whole industries running against each other, some spurred on by government subsidy and an adventurous financial sector, others hamstrung by bureaucratic delays and cautious bankers.

Because biotechnology is such a heterogeneous area of research activity, the professional lobby for biotechnology in the UK is an alliance of a whole range of scientific societies. Its leading lights are constantly engaged in a struggle to get more resources for their collective project. They evidently feel that the whole thing might fall apart, say with a wholesale departure of researchers overseas, a few spectacular commercial failures with new products that would shut off investment, and devastation of the research base by the continuing effects of economic decline. Because of this the British proponents of biotechnology rarely breathe a word about its possible negative aspects. Surprisingly perhaps, in America, where there seems to be much greater confidence that biotechnology will happen, that reticence about problems is far less common. Just as there is faith that entrepreneurs will get the products into the marketplace, ahead of, or in spite of, the competition, so also there is some publicly expressed optimism, "interested" and partisan optimism to be sure, that the problems engendered by entrepreneurial activity can be managed. More aggressive forms of "technology transfer" from the university or government institutes may create strain, but it can be dealt with.

Conflict of Interest as a Manageable Problem

Ever since 1974, scientists and others have remarked with satisfaction that public debate has been more constrained and sensible in Britain than in the US. No British university has debated the hazards of recombinant DNA research; no prestigious scientific society in the UK has organised a public symposium on the social implications of biotechnology, as occurred in Washington and Amsterdam. No British parliamentary committees have debated the questions of changes in patent law and the impact of commercial involvement on academic research. All these things have happened elsewhere, principally in the United States.

Far from seeing this as a victory of British common sense over American controversialism, I see it as a cultural disaster. The negative impact of such debate on US research has been minimal and its positive effects real and enduring. An American historian of science, Charles Weiner, has argued that such discussion, far from acting as a brake on scientific progress, has in fact served to accelerate it by bringing it much more rapidly to the attention of those who hold the purse strings than is usually the case for emergent research fields. Similarly the bureaucracy established to oversee research, the Office of Recombinant DNA Activities at the NIH, far from being costly, irrelevant and cumbersome, as some scientists allege, has, according to Weiner, facilitated communication about research methods and a useful standardisation of procedure. Furthermore in the United States, public debate, some of it caricature, fantasy and phobia from both "sides," and some of it highly informed and rational, has served to get implicit assumptions scrutinised and dogmatic beliefs revealed as such. America comes out of all this as a society more committed to, and more at ease with, adversarial debate, more used to staging conflicts as a means to their resolution rather than pretending they are not there. At the end of the day formal codes of conduct and institutional policies may get violated just as unspoken rules of good form and decency do, but at least potential problems are brought to the surface earlier and the institutional means to get them thrashed out are stronger and better known.

An example may illustrate what I mean. It concerns the establishment of an institute for research in the life sciences at the Massachusetts Institute of Technology. Edwin C. "Jack" Whitehead made a fortune from his biomedical equipment company called Technicon, which sold automated blood analysers, among other things, and made considerable profits from them. In due course Whitehead sold his company to the cosmetic giant Revlon, thereby gaining shares in Revlon and consolidating a large personal fortune. He still has business interests, which we consider in a moment. Throughout the 1970s he sought to endow a research institute at an American university, but for reasons that are not clear this proved difficult to achieve. At one point he almost reached agreement with Duke University in North Carolina, but the deal was abandoned. Nobody has yet explained why it should have taken so long to find a university willing to accept the $125 million that Whitehead has now offered MIT.

Because of these difficulties, Whitehead hired Joshua Lederberg (Nobel prizewinner for his work in molecular genetics and now president of Rockefeller University in New York) to find someone willing to run an institute of the kind that Whitehead has in mind. It is just a little surprising that Nobel prizewinners are available for that kind of headhunting, but *noblesse* is nothing, I suppose, without the power of patronage. Lederberg himself was into commercial biotechnology very early on as one of the founders of Cetus Corporation, while he was still on the West Coast. Through Lederberg, Whitehead came to meet David Baltimore, Nobel prizewinner for his work on tumour viruses, and a professor biology at MIT. That meeting led to an agreement that Baltimore would head Whitehead's institute, wherever it was to be located. Various universities, Stanford, Rockefeller and Harvard among them, seemed eager to accept the proposed institute, no doubt because of Baltimore's association with it. In the end it has gone to MIT. The announcement of the planned links with the university created a storm. Initially MIT faculty were forbidden even to discuss publicly such details of the Whitehead proposal as were known to them. Gradually however more information has come to light, leading to a full-scale debate by the MIT staff in November 1981, which eventually approved the basis of the planned donation of $125 million.

Three issues have come in for critical treatment. They are, first, the independence of an institute, associated with MIT and gaining from its prestige, but not formally controlled by it. Second, there are potential conflicts of interest that could affect the director and researchers at the institute. Third, some critics claim that it represents a takeover of a fundamental research area by private capital with no real safeguards of the public interest.

One hundred twenty-five million dollars buys a lot of bricks, steel and glass, a great deal of scientific equipment and a surprising number of people. The intention is that the Whitehead Institute should have 200 staff, of whom thirteen would be on the faculty at MIT, although paid from the Whitehead budget. Some people have seen real problems of power and accountability here. For example, who would actually control issues like promotion, time for sabbaticals away from the university, the amount of industrial consulting or the availability of Whitehead researchers for teaching—MIT committees or the institute director? These may sound mere formalities, but they are vital to healthy social relations amongst highly competitive people. The fear of some people at MIT is that the institute researchers will use the name of MIT to glorify their endeavours without properly acknowledging their debt to the institution.

The second point follows from the first, but has to do with money rather than status or obligations to teach or to supervise research. Edwin C. Whitehead is in business as a venture capitalist and has investments in at least two biotechnology companies, Plant Genetics, Inc., and Liposome Co. Liposomes are small globules of fatty materials (lipids) that are used to package biological materials, such as antibiotics, chemotherapeutic drugs or even DNA, to get them delivered to a particular site in the body, without being attacked by marauding enzymes and antibodies. It is a technique that may have important applications in chemotherapy, particularly for cancer, and in genetic engineering. It has already been patented.

On the face of it, there is no obvious connection between the commercial exploitation of liposomes and plant genetics and research in developmental biology, which is to be the focus of the Whitehead Institute programme. But such connections could easily appear, and critics have claimed that they could be a powerful influence on supposedly disinterested research. Under those conditions the Whitehead Institute would become an unacknowledged research wing of Whitehead's companies. Baltimore claims that this will not happen. The problem, if it ever appeared, would be solved, he claims, by Whitehead selling the companies involved.

Perhaps as important are the commercial involvements of the staff of the Whitehead Institute themselves, since they are much more closely associated with the actual research going on. David Baltimore has a major investment in a company called Collaborative Research of Waltham, Massachusetts, which is backed by the Dow Chemical Company. It is said that this fact was unknown to some of the MIT faculty debating the Whitehead issue, and that had it been known then more concern would have been expressed about potential conflicts of interest.

Apparently it is the Collaborative Research connection that worries people most. Baltimore consults for the company and owns 300,-000 shares in it that are worth about $3 million. His MIT colleague David Botstein has spent a sabbatical year there. So too has Gerald Fink of Cornell University, who may join the Whitehead Institute. As one research administrator puts it:

> They have a potential conflict of interest here. At least in theory, MIT can't be sure that the decision-making over whom those patents will be licensed to won't be made by people with vested interests in seeing that those patents go to certain companies.

The implication is that the benefit of research done at MIT might be channelled preferentially and at MIT's expense to companies in which Whitehead researchers have a commercial stake.

Some people see these problems as being easily solved by a supervisory committee within MIT and a policy that requires public disclosure of commercial interests. Stanford University president Donald Kennedy has been trying hard to increase its revenue from the industrial use of its research, and accordingly they have had to think through many of these problems. Stanford has, for example, a new set of rules on "tangible research property," such as cell-lines, inspired by the dispute between UCLA and Genentech over the use of interferon-producing cell-lines. There is a committee on research looking at faculty consulting, outside employment of graduate students and the like, and there may be rules that prevent Stanford University from co-investing with a faculty member in his or her business enterprise.

This question had arisen when Harvard University considered placing money in a company being set up by one of its biology professors, Mark Ptashne. This proved an extremely controversial move, and the proposal was voted out by the university staff. They felt that the allocation of resources within the university could all too easily be influenced by commercial pressures owing to the particular directness of the link between the research of a team on campus and the financial fortunes of the university as a whole.

However, universities are likely to invest more indirectly in biotechnology. For example in the UK, Advent Management Ltd. is investing funds from Oxford and Cambridge in new technologies, and in due course genetic engineering will form a part of these. At Stanford, something of a pioneer in university commercial involvement, there is a rule requiring disclosure of industrial connections on request. This is thought to be sufficient to prevent some of the uglier problems of conflicting interest. Britain and America share the general opinion that these problems can be managed, and that guidelines can be policed. In the US, however, public discussion is actually believed to aid this process. In the UK, conversely, there is simply not the enthusiasm to raise such distasteful truths in public, to admit that people do try to rip off their employers, sponsors, colleagues and subordinates. Nor is there any desire to work out some institutional controls that would make such practices much riskier.

For example, a so-called BioCentre at Leicester University has recently been set up with money from four private concerns. ICI jointly operates a laboratory there already. However, so far as I am aware the Leicester proposal has provoked no great debate. It may not be on the same scale as ventures in America, but it is a remarkable step on to the campus and a taste of things to come. It would be amazing if the sudden appearance of this kind of investment did not create all kinds of strains, yet the institutional procedures for grappling with them in British universities are very primitive.

It is difficult to have confidence in the ability of British universities to regulate the behavior of their senior members if and when the suspicion of gross irregularity is raised. Although I am not trying to present American institutions as necessarily less corrupt, one can't help being struck by the readiness to discuss these problems openly. The British are too imprisoned by a myth of decency and financial integrity in public life.

Even with clearer guidelines, another major problem remains, and it is one that, although mentioned in the debates at MIT, and present in many other situations, has tended to exercise people far less. After all, everyone could behave with perfect propriety, consulting outside the university for no more than one day in five, disclosing every Concorde flight and night at Claridge's with meticulous care, and something still be wrong. The difficult question is how the exploitation of publicly funded research ought to occur. This question came up at

MIT, as it has on various other occasions when research teams, their skills and plans have been bought up by industrial concerns. Is it just or healthy for research in the medium term for that kind of appropriation to go on?

Let us be clear about the shape of such arrangements. Grants of say $50 million are being made to university researchers by industrial corporations to fund several years research by an established group in return for an exclusive right to patent and commercialise any ideas that emerge from the research. The advantage to the corporation is that it gets an already established, functioning, productive group that is on tap at bargain prices as a source of product ideas. No new facilities need be built, no additional training paid for. All that has to be provided is the money for research. The likelihood is, moreover, that the university people will experience minimal strain as the harness is fitted over them, and that they will be eager to show that they are worth the money. They stay on campus, close to friends and colleagues. They become bigger fish in a small pool rather than the boffins in the research department. The advantage for the university is the injection of funds at a time when government funding is getting harder to set up. They charge for overheads, and can attract staff eager to get into this kind of relationship.

But there are also costs to the university, which cedes some power over the control of research strategy, and to the taxpayer, whose taxes have financed the constitution of research expertise, now being directed to the advantage of the new corporate patrons. In some cases the first objection has been countered by the establishment of university-sponsor committees that are supposed to review the impact of this funding on research. It remains to be seen whether in fact such committees will be able to perceive any bad effects, such as secrecy, disrupted communication, priority disputes, changes in citation practice, or a shift in research programmes away from the ideas with a longer-term payoff or which conflict with industrial interests. I doubt whether such scrutiny will mean very much, since at the end of the day what will matter is getting the next share of money. Those with the money will be able to buy pretty much what they want.

The second objection concerns grabbing the fruits of expenditure from the public purse. Corporations pay taxes, a tiny proportion of which is spent on academic research as a general social investment in invention and training. When researchers come up with something new, industrial concerns may step in and buy the right to develop that idea and accept thereby the commercial risk involved. From this perspective anything that speeds that process is self-evidently a good thing. Why erect barriers to technology transfer, when universities get an economic return on their inventions through the patent system? The return to the taxpayer comes when the firm pays taxes on its profits. I think even in these terms there is a problem. Academic research would be taken over very cheaply and it would amount to a form of general social subsidy to privately owned industrial corporations. That in itself might be sensible, if industrial innovation was aligned with real social needs. As I have tried to show that is not always the case.

The real problem with industrial involvement in fundamental research, in my view, is not only that skills and ideas are being obtained at too low a price *but that the possibility of alternatives is shut down.* In David Noble's phrase, the ivory tower is "going plastic"; it is becoming a think tank for major corporations, which have exclusive access to it, rather than remaining a general social resource. This is not an argument for the ivory tower *per se.* Isolation and pure academicism seem to me as undesirable as complete subordination to corporate capital. It is an argument for a measure of university independence, guaranteed by the state, to maintain a plurality of ideas, critical thought and a vision of social, economic, industrial and scientific alternatives. This then is why the question of conflicts of interest is either trivial if it refers merely to observing the rules and not being greedy, or significant when it points up the fact that the value of research is not simply its value to capital. If its value comes to be seen exclusively in those terms, then something vital has been lost.

Technology and the World Economic System

In the three central chapters of this book I was trying to set technical development in biotechnology in context, to present it as part of an economic system, as devices for the continuation of a particular mode of production. The medical fruits of gene-splicing fit a certain kind of health care: hospital-based, curative, highly specialised, and profitable for the companies that service it. Pharmaceutical companies have achieved their present power because of their ability to package curative materials, to set their prices and to train doctors to think in drug-oriented terms. What the general public gains from this is a service with a highly developed ability to cope with conditions that lend themselves to mechanistic care. What we lose is any real concern with preventive medicine, with illnesses that afflict those held to be no longer productive in our society and with complex states like anxiety and depression.

The losses in less developed countries are different. Much Western medicine is simply too expensive to be of use there. When exported it is either useless in different social and medical conditions, or helps only a tiny minority who are free from the crippling effects of economic dependence on the wealthier nations. That is the background that has to be kept in mind when thinking about what biotechnology means.

The economic forces that have led to the constitution of a certain kind of medicine operate continuously on medical biotechnology. It is their latest offspring, their project, their hope for the future. It is a route into new markets, selling a kind of health care that has proved efficacious and profitable, appealing and costly. It is a way of continuing to do business by making molecules that can be sold at a profit. Health, defined here as the absence of clinically significant symptoms, can continue to be produced by the administration of curative agents: antibiotics in the 1950s, psychoactive drugs in the 1960s, prophylactics for heart disease in the 1970s, human insulin, interferon and monoclonal antibodies in the 1980s.

Similarly with food and agriculture. I don't think we can make much sense of what is going on without a feel for the dynamics of industrialised agriculture and the plans of the corporations that service it and process its products. Just as modern medicine offers a remarkable array of curative tactics within a strategy that leaves the structural causes of disease alone, so too agriculture leaves the fortunate few well provided for, while the global structure of production denies adequate resources to millions in need. Just as we have adequate medical knowledge to prevent and treat a great deal of disease, if the necessary resources were distributed equitably, so too many more of the world's population could be fed adequately if resources were distributed differently. As it is, far too much land is given over to the cultivation of crops that people can't eat, and the revenue from those crops, if it ever returns to the country where the rubber or the cotton or the coffee grows, accrues only to a tiny social élite. Far too much effort is devoted to persuading peasant farmers to use inappropriate systems of agriculture which make them dependent on expensive imported fuel, seeds, fertilisers, pesticides and equipment. In the developed world farmers with far greater resources at their command but with this same dependence on fertilisers and cheap energy are encouraged to overproduce, and some of the resulting food surplus is used, not as aid, but as a weapon to ensure compliance with the economic and political interests of donor countries. That then is the background to much agricultural biotechnology, making agribusiness more profitable by maintaining farmers around the world in dependence on its products.

The chemical and energy sectors fit the same picture. Energy corporations are among the most powerful in the world. Their turnover is greater than the gross national product of many countries. They have the power, not only to defy governments, but occasionally to change them. Selling petroleum is very big business indeed. It has proved an effective base from which to diversify into other areas, notably chemicals. That sector too has its multinational giants, squeezed at the moment by fundamental economic problems but still capable of mustering immense resources to restructure around a different starting point. Whereas the costs of pilot plant for coal gasification might run into hundreds of millions of dollars and be written off by an oil company if necessary, in chemicals tens of millions are ventured on new schemes that may also never achieve economic viability. At the end of the day the idea is to find a hydrocarbon feedstock that, with a minimum of energy, can be elaborated into far more complex molecules, including those for sale in the agrichemical and pharmaceutical sectors. The plan is to restructure a global industry around a new starting material and energy source in ways that lead profitably to a myriad of plastics, solvents, resins, paints, fertilisers and fine chemicals. The hydrocarbon economy will have a different foundation, a slightly different geography but the same commercial basis. Large producers using highly automated, centralised process plant will attempt to sell enormous quantities of goods produced by two increasingly separate armies of labour, the engineers, corporate planners, sales people and production management on the one hand, and unskilled labourers on the other.

Where Do We Go from Here?

The preceding four sections offer partial but important views of biotechnology. They offer perspectives on what is happening. None of them can be dismissed out of hand; none of them can stand alone. The first is a general comment about the politics of technology. The second may seem familiar, as a litany of complaint about British economic performance. Whether biotechnology is being held back in Britain I can't say. Nonetheless the fact that historically British investment in technology has often been misguided and too late is a serious issue.

The third section is really a story of the potential problems of "success" in fund-raising for research. Whitehead has put up a vast amount of money, and an institute is rising from its foundations. But only the naïve or the disingenuous would deny that the scheme has the *potential* for creating immense problems over what kind of community a university is supposed to be, or whether it is likely to be one at all, in the sense of colleagues sharing a common allegiance, communicating freely and taking on the various tasks necessary to maintain an institution of learning dedicated to the common good. The irony of the Whitehead case is that donation has supposedly been made in a spirit of disinterest. The issue has become controversial only because this claim has seemed mildly implausible to some people.

There are now perhaps twenty or thirty financial arrangements between corporations and universities in the US that specifically concern biotechnology, and where the intention to profit commercially is clear, unambiguous and open. Yet they have excited far less controversy. The arrangement between Hoechst, the German chemical company, and the Massachusetts General Hospital in Boston involved some $50 million. That is a tremendous amount of money for a university research programme. Even if the conflicts of interest can be "managed," it is still a remarkable *Anschluss* of academic territory with the corporate goals which it now serves. For, crude and dogmatic as this may sound, biotechnology is being penned and shackled so as to serve such global masters. The biggest chemical, oil, fertiliser, food, seed and pharmaceutical companies are the dominant actors in shaping the future directions of this field. Their interests are the maintenance of profit and power through the sale of commodities. These address certain needs, which are defined in ways that such corporations go to immense lengths to specify, influence and control. Their interests as corporations are coming to act on and through biotechnology in more and more direct ways. The purposeful seeking out of opportunities for profit is being built ever more intimately into the form of the scientific enterprise.

Analysing programmed material is more and more frequently the preliminary phase of product development. Remember the patent lawyers in the Genentech seminars, shaping ideas for their ownability as soon as the vocal cords and lips release them. Remember the erstwhile colleagues in San Diego, manoeuvring around each other in noncommunication, knowing even before anything is in print that synthetic antigens could earn them a lot of money, if they can establish a prior claim on the ideas involved. Remember the controversy over the Agrigenetics patent where a technique said to be in constant use for breeding new plants is suddenly whisked away as private property, its new owners claiming that the particular way in which they connected together all the constituent ideas entitled them to ownership. Increasingly the term "technology transfer" is losing its meaning, since there is no motion or transfer. Conception *is* capitalisation.

So where do we go from here? Must we just sit and watch the juggernaut gather speed or, as I put it, in rather reflective mood, at the beginning of the chapter, enjoy "a rare historical opportunity to watch it all in motion, to see the technical skills of gene-splicing being used to unlock another cycle of industrial change"? Or could we try to construct and implement different technological and industrial strategies that frame the possibilities of biotechnology differently? Could we change the priorities of the research now going on, in the hope of reforming biotechnology in the early developmental stages? I am not very optimistic that we can.

To try to rethink the rationale of such a rapidly expanding field is almost completely crazy at the moment, when so many people see high technology as salvation and a path back to employment and economic growth. When the "iron laws of the market" have been brought into play with such deliberate and effective force to reinforce class discipline on the army of labour, it scarcely seems rational to speak of reorienting one of the technologies into which investment is pouring, or to take a different path out of the slump.

I expect to be told that the realities of these times are that innovation is a political issue only when it slows down or if it fails to happen. The only question of interest to many people is how to get the gene machine up to speed. Even some of those whom one might expect to be interested in the politics of science and technology, like research workers and technicians in white-collar unions, seem to regard biotechnology very pragmatically and uncritically as mere transport to a better future. I can't deny these things, but I can try to change them by, as I put it earlier, "excavating under the rhetoric" and pressing for more consideration of what biotechnology represents as an industrial, political and cultural phenomenon and what the consequent losses and gains will be. Accordingly, my suggestions for a political response to this phenomenon are all intended to be tactical moves, that might help to build greater social concern about what's happening and a capacity to act on that concern.

Trading Participation in Planning for Research Support

First, I think we have to acknowledge the fact that biotechnology, like any other form of radical innovation, like anything that is more than mere fiddling with existing products, requires promotion, government support and the creation of a research base. In contemporary society, innovation is a race.

Biotechnology involves two kinds of expertise, one involving molecular genetics and the reprogramming of biological organisms, the other concerned with chemical engineering and the design of reactors within which biological materials are transformed. Throughout the book I have stressed the first rather than the second. Both are clearly important, as many research companies founded on virtuosity with bacterial genes are now discovering. What counts as "excellence" and who sets the standards are the first things to consider. Second, the allocation of research funds by committees of leading scientists is usually a secretive exercise, with the general principles of policy and the details of specific decisions known only to a few key individuals, often themselves receiving massive support from the same programme. The deliberation over how to allocate funds could be opened up enormously and made far more participatory. Also, the presumption at the moment is that successful research will be commercialised by private companies. One possibility is that specific provision could be made for research aimed eventually at other kinds of productive enterprise, like research cooperatives, interested, say, in the processing of urban wastes.

At the moment, in Britain the research councils, the source of virtually all government money for fundamental research in this area, are also encouraging applicants for grants to form joint ventures with industry. One future condition of such arrangements could be that the company concerned agrees to allow and assist a review of its research and investment plans by trade unions representing its work force. The aim would be to allow workers in that company and the researchers, first, to see where that research was intended to fit into the plans of the company and, second, to make possible the formulation of an alternative corporate plan, which would also involve biotechnological research. The intention here is to make research strategy more visible and negotiable.

These proposals are based on the view that corporate involvement in the development of biotechnology is inevitable, and, at the moment, essential. Whilst publicly owned enterprises could and should play a role, private industry has to be involved, but the terms of that involvement could certainly be changed. In effect, this is a way of buying a greater degree of worker participation using research as the currency.

A similar attitude to university-industry collaboration seems sensible. As a historical trend, it is one that would now be difficult, and perhaps undesirable, to reverse. Just at the moment there are skills in the universities that some companies feel they need. Simply to block access to them would, I think, serve no useful purpose as things stand, and would be immensely unpopular. If corporate funding of university research does expand, we need to ensure that the conflict of interest problems are fully discussed, that technicians, students and colleagues are not ripped off and that effective rules of conduct are developed and enforced. Strong rules on full disclosure would at least help to reveal who was working for whom, on what and for what return. It might make clear why particular individuals were especially keen to open up new research areas, and give greater indication of when academic independence was likely to have been compromised. Where scientists are involved in setting standards or regulations, it is important to know, for example, whether they just happen to be a consultant to a company likely to be affected by that decision.

But the problem of people's judgement getting bent by commercial relationships is only one issue. We have also to consider patent policy and the use of revenues from licensing agreements. For example, is constant vigilance and recourse to the patent system the only option for academics involved in biotechnology? I suspect it is. Advocates of patenting represent the practice as a form of protection for the inventor. In one sense that is obviously true, if and only if one can afford the costs of defence when patent rights are challenged, perhaps deliberately, by a powerful company. Many patent holders, after all, sell their patents for a lump sum, because that is a low-risk, medium-gain option. Protection is available through the patent system, then, but in a more limited form than many people realise.

This takes us to the question of who should hold patents and to whom the resulting revenue should return. The Cohen-Boyer patent is an interesting example. The revenue goes to two universities and not to the individuals. The patent cover was arranged by the universities concerned and not by a government body. It was a local initiative to subsidise a general programme of research. It is in effect a tax on commercial recombinant DNA research, that will be passed on to consumers, and it will benefit research work at Stanford and UCSF alone. Many people have felt that the contributions of researchers at these institutions to the emergence of gene-splicing were not such that they should have been privileged in this way.

There is another option for the protection of publicly funded research against wilful private appropriation, and in principle I find it a more appealing one. The state can seek and hold patents on such research. This is in effect the British system. The National Research and Development Corporation, now part of the British Technology Group, has an exclusive right to license the commercialisation of publicly funded research. It has had its successes, notably the cephalosporin drugs through which it has earned hundreds of millions of pounds, but it has also had its failures. Some people would say that the case of monoclonal antibodies was one of them. Its critics say it is too slow, too inflexible, too uncommercial and too demanding of potential licensees. This is blamed on bureaucratic failings and its monopoly position as a broker of ideas. They may be right, although I suspect that kicking the NRDC is a convenient target for academics with no real idea of how difficult licensing and commercialisation are. Perhaps a less centralised regional system, like that in France, would work better. Clearly if the principle of government licensing in the public interest is to be retained, and I believe it should be, then the alleged shortcomings have to be made public and, if the allegations are justified, put right.

What should happen to licence fees for university research? One opinion is that research groups within universities should be run, in effect, as private companies and given the lion's share of any revenue. I think it is preferable to try to maintain the independence of university researchers by dissociating their work from such a direct connection with the market. This can be accomplished to some extent by running a research trust fund into which commercial income is paid. One example is WARF (Wisconsin Alumni Research Fund) at the University of Wisconsin, which was established to absorb money coming in from patents on vitamin D. WARF also gave the world the rat poison Warfarin. One possible disadvantage of this arrangement is that with financial benefit can come legal liability, as WARF has recently discovered. Universities can settle for rather less revenue by selling technology licences for a commercial company, and acquire immunity from prosecution thereby if anything goes wrong. All these schemes will have to be examined by research organisations as biotechnology goes commercial. Whichever is chosen, it is important to analyse its losses and gains periodically and to do so in the public gaze.

But licensing can't happen without licencees. What we have also to consider is what kinds of institutions should commercialise biotechnology. Should we hand over this new industry to private enterprise, or should we try to ensure that it remains in public ownership? The success of the entrepreneurially directed research companies in the United States seems to have given many people the impression that innovation in this area can only be promoted by cult heroes using private risk capital. Mrs. Thatcher's visit to Genex, where she enthused about permissive US tax laws to help the small entrepreneur make money for those with spare cash, was clearly an endorsement of that form of innovation.

At the moment decisions about research funding are carried out in a very exclusive way. It would be good to see far greater lay participation in this process. Under such scrutiny research proposals that actually call into question doctors' expertise, or seek to transfer power from doctors to patient, or which focus specifically on the social causes of disease, like hazards in the work place, might fare better than at present.

Finally we need to think about risk. Many scientists will say that the risks of recombinant DNA research were overblown. Exaggerated fears of new infectious agents and ecological disaster were, they would say, only contained by the most vigorous programme of opinion management, which led to the cutting back of rampant regulatory bureaucracies that would have stifled the new field of research.

I still find it amazing that biologists are prepared to make such definite statements about the safety of recombinant organisms after only ten years work with tiny quantities of them. It is not that I think they have missed something already. I just don't understand how some people can be so certain that the biological future holds no nasty surprises. Accordingly, I think the risk assessment of biotechnology must not be allowed to die. Production with large volumes of more robust micro-organisms raises new biological questions. National advisory bodies like GMAG in the UK need to review their knowledge of large-scale biotechnology continually.

I see all the measures suggested so far in this section as defensive, designed to hold the line against the further incursion of corporate capital into the biotechnology laboratory. I am not arguing for an ivory tower, or for regarding nonapplied research as a loftier, more noble enterprise. My intention is to identify the sort of measures which would get popular support, which would not be economically or scientifically disastrous to implement and which would allow for more radical structural changes in research and industry to come about. At the very least I would like to make clear that there are ways in which the priorities now being built into biotechnology could be examined critically, and new, more participatory, more democratic institutions devised to shift the way in which research is conceived, appraised, funded and managed.

Putting Biotechnology on More People's Agendas

In certain circles the financial, industrial and political implications of biotechnology are receiving a great deal of attention. In boardrooms, senior common rooms, stockbrokers' offices and the corridors of power these things are a matter for focused thought. All this makes a powerful contrast with the relative lack of concern shown in biotechnology by trade unions, nonindustrial pressure groups, consumer groups and representative institutions of our society, outside the central or federal government agencies.

One of my purposes in writing this book has been to try to broaden interest in the idea of thinking through what the implications of biotechnology might be, since at the moment few of the technological choices have been set in concrete. This kind of anticipatory exercise is all too rare with new technologies. But at least with micro-electronics there are some signs that a sense of their probable impact on work, employment, leisure, consumption, education, expertise and the division of labour is making all kinds of people say, "What will this mean for me—and my job/educational career/self-esteem/social status/personal relations?" Something similar must begin for biotechnology.

To this end I want to suggest what kinds of constituencies could start to ponder the possible futures, and to indicate the kinds of questions that they might pose. Essentially then, those reading this book should reach this point and ask themselves now, if they haven't done so before, "What is all this likely to do to my job, way of working, domestic activity, health prospects, urban environment, power as a consumer and user of energy?" They might also ask, "How could this be taken up by the professional, civic and voluntary organisations to which I belong, so as to gain more influence or control over what is happening? In what ways could those organisations politicise the process of innovation?"

For example, if the chemical industry changes over to fermentation processes it is important that the possible hazards of working with micro-organisms like yeasts are fully considered. If urban waste becomes a valuable commodity, what would this mean for the work and wages of municipal sanitation workers? For farmworkers, it is important to consider how new plants are being designed into a new form for the industry. Will they continue the trend to mechanisation, to larger farms, and to less skilled work? Generally workers could consider whether biotechnology could be used to create more effective forms of occupational medicine.

In the field of health and medicine, the dominant question for the foreseeable future is likely to be what kind of health service will exist and how much of it will be in public ownership. It is not obvious that research policy will have much effect on the outcome of this struggle, but more resources for preventive medicine would, in the long term, allow less burdened hospital staff to give a better standard of care to their patients.

Vaccination is one form of preventive medicine that could be effected by biotechnology—e.g. for dental caries—yet in this case a more important form of prevention would be if people changed their eating habits and became more aware of how to look after their teeth. This is the kind of strategic question that will be dismissed within the dental profession. Dentists could do more to spread these debates beyond these narrow confines. Similarly, contraceptive vaccines might sound like a good idea, but experience with the controversial injectable contraceptive, Depo Provera, has shown that it is often used covertly without the women concerned being consulted.

One of the problems with modern medicine is that it is carried on in large impersonal institutions, the logistical and organisational needs of which often come before the medical and psychological needs of patients. Women's groups particularly have pressed for a different kind of health care, to eradicate the sexism that is deeply engraved in much medical practice and theory, and to transfer power of decision-making away from the doctors and back into the hands of people seeking medical assistance. In chapter 4, I mentioned rather briefly that monoclonal antibodies could make sex predetermination possible by allowing selection at some point in the process of procreation between male and female sperm. If this becomes technically feasible it could have serious effects in reinforcing and amplifying discriminatory and oppressive attitudes to women, who would grow up with the experience of having been chosen after their older brother.

To try to counter this process it is important to think how women could use medical developments to support confidence in their abilities and to shift power away from the managers of the clinical team. One possibility also concerns monoclonal antibodies, which offer the prospect of more accurate, cheap pregnancy testing. While this is often a commercial or hospital-based practice, it could be carried out by a community-based women's group, able to provide a different kind of advice, support and access to other services than existing institutions allow.

In the field of food and agriculture there are all kinds of questions to consider. One of them is the loss of plant species taken up by the World Wildlife Fund to redress the balance of interest away from animal species. This seems to me an issue that consumer organisations could consider. Other bodies like the British Soil Association have an interest in what plants are grown, and how they are grown and harvested. Organisations concerned with poverty, hunger and exploitation in less developed countries like War on Want and Third World First could analyse what gasohol programmes and new forms of agriculture are likely to mean.

There is in fact an amazing amount of forecasting, speculation, trend-guessing, scenario-building and discussion that could go on. Once you start to see the possible ramifications of biotechnology, and the range of groups in society whose interests, status and powers could be either advanced or destroyed by the coming revolution, the prospect of a self-catalysing social debate about the goals of this technology is both exciting in prospect and urgently necessary. It makes you see how much goes undebated at the moment.

A Culture of Futures

You cannot write a book like this without speculating, extrapolating into the future and making what you hope are informed guesses. Sometimes, I am sure, I have got the trends wrong or overlooked what now seems to be a minor phenomenon, which will, in ten years time, come to dominate the scene. I feel some responsibility for trying to minimise the looseness of fit between this text and the future. Futurologists must have standards, but there are limits to what can be done. Where it isn't clear what will happen, I have tried to indicate what I think is the likely pattern of development and to display my uncertainty. I also hope that I have made clear how I tend to approach the task of forecasting, from a critical view of the economic enterprises that drive technological change forward. I do think that biotechnology will have some unambiguously positive benefits. I also think that its developments will be shaped and structured by a powerful combination of research companies and large multinational corporations (pilot fish and sharks, maybe?), to bolster a system of production that I find horrific even though I am one of its more privileged beneficiaries. I also think that the way goods and services are produced could and must be changed, so as to decrease, and, in the long, long term, abolish the inequality, poverty, exploitation, brutalisation and emotional impoverishment to be found in the present system.

I would like to see biotechnology used to begin that process, and to build the confidence to do things differently. I wrote this book out of a sense of outrage about the way things are going, and I hope it shows.

The problem is that the social confidence to respond critically to science and technology is very low at the moment, just as the pouring of the foundations of new science-based industries is taking place. How frustrating and sad it is that these critical phases of technological innovation, when the basic elements of the industrial infrastructure are redesigned, come at moments of political collapse and disintegration. When dissent, scepticism, appraisal, lobbying for alternatives, political debate and public rumination are particularly necessary in the field of science and technology, the forces and groupings that might promote this task are driven backwards, forced into compromise and broken. At the moment, this kind of retreat by labour and advance by capital is also the background for the assiduous courting of academic scientists, whose expertise is presently, and perhaps only temporarily, necessary to help this corporate thrust into the future on its way. I have already indicated one possible means by which that expertise might be reoriented, by linking funds for research with political reforms that would alter the way that industrial investment is planned and negotiated. But these things will not happen unless other social forces are let loose, and cultural attitudes to expertise, to scientific and technical competence and to academic knowledge are changed. Scientists' goodwill, integrity, responsibility and far-sightedness—and all these reflexes are being dulled anyway by the fall in the economic temperature—are not enough on their own, even though their spokespeople will no doubt continue to say that they are. Without a more general ferment of questioning, very little will change.

Earlier in this section I mentioned "futurology," a slightly dubious claimant to respectability as an academic discipline, but an interesting area nonetheless. Futurologists try, as their name implies, to divine the shape of things to come, the ordering principle, or *logos,* of the future. They aspire to be a discipline like psychology or archaeology, and that is a noble aim. It implies the working out of shared intellectual standards to separate the wheat from the chaff, a set of techniques to learn, master and pass on, and a coherent intellectual programme. On the other hand, assuming the form of a discipline is also a kind of retreat from the public, a turning inwards usually expressed in language as a technical community starts to mumble to itself, ceasing to think about the puzzled, once-interested onlookers. Disciplines are then both a possible route to intellectual rigour and a terminological maze in which the uninformed soon get lost and turn back for the entrance in panic.

Could we perhaps, as a society, construct a futurology that had some rigour and reliability without the closure of an academic discipline? I am convinced we must try, since the rate at which the future is coming to meet us is increasing, whilst our ability to think through the implications of interacting scientific, economic, political and cultural trends remains very underdeveloped. I am not talking about sensational, headline-catching gabbling about possibility. I am not talking about esoteric economic modelling or contract scenario construction for private corporations. I have in mind a culturing of possible futures to imagine what they would feel like. The hope is that we would learn to adopt a less passive attitude towards innovation and start, first, to interrogate technical experts in a more confident way, and then to participate in the process of designing the future. Fine sentiments, but what could be done to promote this kind of cultural change? To talk of holding something in the public gaze is, of course, to talk of the mass media—wall posters, books, magazines, radio, television, cinema and video cassettes. I think we have scarcely begun to use the media to respond imaginatively and critically to science and technology. Certainly there are now more analytic programmes about micro-electronics, manufacturing and employment than there used to be, but that is only a small part of the technological globe, albeit an important one. Then we also need to learn to question the available experts, particularly about the assumptions on which their judgements rest and what model of society, production, wealth creation and politics they are employing. It is commonplace to tackle politicians aggressively on the media, though they may walk out of the studio. Scientists on the other hand are treated far more reverentially. I am not suggesting that we should start throwing scientists to the lions as a kind of televised Roman circus to entertain us as pure spectacle, but I do think we should press experts harder on the reasoning behind particular scenarios and policies. After all the future depends on it.

Notes

CHAPTER 1

page 1 "into this complex . . . " This point is well brought out in the issue of *Scientific American* for September 1981 on "Industrial microbiology." This is an excellent introduction to the field.

page 2 "Yet already the first wave . . ." See R. Hardman, "Top biotech firm says no to Britain," *Sunday Times Business News* (21 February 1982); "Growth bug," *The Economist* (18 September 1982); "Gestation blues," *The Economist* (31 July 1982); "A shaky start for genetic engineering's first bug," *The Economist* (25 September 1982); D. Fishlock, "A jaundiced eye on bioriches," *Financial Times* (25 August 1982). See also, J. Elkington, "Signs that the biotechnology bubble is about to burst," *The Guardian* (1 April 1982), 19.

page 3 "For example take human milk . . ." See H. Marcovitch, "Mother's milk: beware of imitations," *New Scientist,* 88 (6 November 1980), 371–3. This shows the complexity of human breast milk and the near impossibility of simulating the action of the breast in lactation. For the serious health problems associated with the use of artificial milk substitutes, see L. Doyal, *The Political Economy of Health* (Pluto Press, London, 1979), 128; also M. Muller, *The Baby Killers* (War on Want, London, 2nd edn, 1975). For an example of the scientific work on milk protein genes see N. M. Mehta et al., "Cloning of mouse betacasein gene sequences," *Gene,* 15 (1981), 285–8.

page 5 "Biotechnology then offers . . ." See M. W. Fowler, "Plant cell biotechnology to produce desirable substances," *Chemistry and Industry* (4 April 1981), 229–33. Japanese scientists have already produced tobacco in vessels up to 20,000 litres in size; at the Wolfson Institute of Biotechnology at Sheffield University they can manage 100 litres. The tobacco has a commercial value as cigarette material and its cells also produce valuable diagnostic enzymes: see also "A growth industry based on single plant cells," *New Scientist,* 89 (26 March 1981), 813.

page 6 "Professor Moo-Young . . ." See "Making a meal of wood wastes," *New Scientist,* 89 (23 April 1981), 224.

page 7 "At the moment new bacterial . . ." See "The protein connections," *ICI Magazine* (January–February 1981), 4–8; J. Elkington, "The greatest thing since sliced bread," *The Guardian* (17 June 1982), 17; "A matter of diet," *The Economist* (26 March 1981), 96–7.

page 7 "They chose to . . ." See D. A. Hopwood, "The genetic programming of industrial micro-organisms," *Scientific American,* 245 (1981), 67–78.

page 8 "These kinds of techniques . . ." See G. Melchers, "Somatic hybrid plants of potato and tomato and protoplast function in potato breeding," *In vitro,* 15 (9) (1979), 216–17.

page 8 "In 1981 researchers . . ." See J. L. Marx, "Globin gene transferred," *Science,* 213 (25 September 1981), 1488; see also "More progress gene transfer," ibid. (28 August 1981), 996–7. See also W. F. Anderson, E. G. Diacumakos, "Genetic engineering in mammalian cells," *Scientific American,* 245 (July 1981), 60–93.

page 9 "The theme of . . ." Whilst I dwell on this theme, I do not go very far into the technicalities of gene manipulation. For a good introduction see Jeremy Cherfas, *Man-Made Life: A Genetic Engineering Primer* (Blackwell, Oxford, 1981).

page 9 The quotation is from *Biotechnology: Report of a Joint Working Party* (HMSO, London, March 1980), 16. This report is in turn quoting an article from *The Economist* (2 December 1978). The report was produced by a working party drawn from the Advisory Council for Applied Research and Development, the Advisory Board for the Research Councils and the Royal Society, under the chairmanship of Dr. Alfred Spinks. It is referred to as the Spinks Report.

page 11 "The pharmaceutical industry . . ." See B. G. James, *The Future of the Multinational Pharmaceutical Industry to 1990* (Associated Business Programmes Ltd., London, 1977).

page 11 "Indeed the first product . . ." C. Schuuring, "New era vaccine," *Nature,* 296 (29 April 1982), 792.

page 12 "The food industry . . ." See "Biotechnology: Hoechst aims for versatile process," *Chemical and Engineering News* (21 June 1982), 30; R. V. Smith, "Aspartame approved despite risks," *Science,* 213 (28 August 1981), 986–7.

page 12 "Sophisticated science . . ." See T. Land, "Man-made cereal relieves hunger," *Technology Week,* No. 10 (10 April 1982), 20.

page 12 "ICI has a bacterium . . ." See S. Yanchinski, "ICI to make textiles from bacteria," *New Scientist,* 89 (19 March 1981), 723.

page 14 "Biotechnology is the projection . . ." This argument is developed in greater detail in E. J. Yoxen, "Life as a productive force: capitalising the science and technology of molecular biology," in L. Levidow, R. M. Young (eds), *Science, Technology and the Labour Process: Marxist studies,* Vol. 1 (CSE Books, London, 1981), 66–122.

CHAPTER 2

page 18 "To think of life . . ." A history of biology, written by a distinguished French molecular biologist, that emphasises this point is François Jacob, *The Logic of Living Systems* (Allen Lane, London, 1974).

page 19 "Questions like this . . ." An excellent introduction to this way of thinking is James D. Watson, *The Molecular Biology of the Gene* (W. A. Benjamin, London, 1976).

page 19 "It is an analytic framework . . ." See Simon Pickvance, " 'Life' in a molecular biology lab," *Radical Science Journal,* No. 4 (1976), 11–28.

page 19 "One biochemist . . ." See Erwin Chargaff, *Heraclitean Fire: Sketches from a Life before Nature* (Rockefeller University Press, New York, 1978).

page 20 "This kind of biology . . ." This historical argument is developed in E. J. Yoxen, "Giving life a new meaning: the rise of the molecular biology establishment," in N. Elias et al. (eds), *Scientific Establishments and Hierarchies, Sociology of the Sciences Yearbook,* 6 (Elsevier, Amsterdam, 1982), 123–43. See also P. Abir-Am, "The discourse of physical power and biological knowledge in the 1930s: a re-appraisal of the Rockefeller Foundation's "policy" in molecular biology"; *Social Studies of Science,* 12 (1982), 341–82; and R. E. Kohler, "The management of science; the experience of Warren Weaver and the Rockefeller Foundation programme in molecular biology," *Minerva,* 14 (1976), 279–306.

page 23 "Some historians . . ." See S. P. Strickland, *Politics, Science and Dread Disease: A Short History of the United States Medical Research Policy* (Harvard University Press, Cambridge, Mass., 1972).

page 23 "One school of . . ." A general history of molecular biology from the 1930s to the 1970s is H. F. Judson, *The Eighth Day of Creation: Makers of the Revolution in Biology* (Simon and Schuster, New York, 1979).

page 24 "Several major discoveries . . ." These are described in more detail in R. C. Olby, *The Path to the Double Helix* (Macmillan, London, 1974) and F. H. Portugal, J. S. Cohen, *A Century of DNA: A History of the Discovery of the Structure and Function of the Genetic Substance* (MIT Press, London, 1977).

page 25 "occasionally queried . . ." For an interesting account of a recent challenge and its ruthless suppression see T. D. Stokes, "The double helix and the warped zipper; an exemplary tale," *Social Studies of Science,* 12 (May 1982), 207–40.

page 25 "In recent years . . ." See particularly A. Sayre, *Rosalind Franklin and DNA* (Norton, New York, 1975).

page 26 "James Watson's . . ." His own account of this work is *The Double Helix: A Personal Account of the Discovery of the Structure of DNA* (Atheneum, New York, 1968; also Penguin paperback).

page 28 "Furthermore, feats such as . . ." The media reporting of achievements like this is analysed in E. J. Yoxen, *The Social Impact of Molecular Biology* (Unpublished Ph.D. thesis, Cambridge University, 1978).

page 31 "It would be . . ." In a sense this cannot now happen as there is a growing number of books on the recombinant DNA debate: see R. F. Beers, E. G. Bennett (eds), *Recombinant Molecules: Impact on Science and Society* (Random House, New York, 1977); J. G. Goodfield, *Playing God: Genetic engineering and the manipulation of life* (Random House, New York, 1977); C. Grobstein, *A Double Image of the Double Helix: The Recombinant DNA Debate* (Freeman, Reading, 1979); R. Hutton, *Bio-Revolution: DNA and The Ethics of Man-Made Life* (New American Library, New York, 1978); D. Jackson, S. Stich (eds), *The Recombinant DNA Debate* (Prentice-Hall, Englewood Cliffs, 1979); J. Lear, *Recombinant DNA: The Untold Story* (Crown Publishers, New York, 1978); M. Rogers, *Biohazard* (Avon Publishers, New York, 1979); J. Richards (ed.), *Recombinant DNA: Science, Ethics and Politics* (Academic Press, New York, 1978); N. Wade, *The Ultimate Experiment* (Walker, New York, 1979); S. Krimsky, *Genetic Alchemy: A Social History of the Recombinant DNA Controversy* (MIT Press, Cambridge, Mass., 1982). In addition, Susan Wright, who is also preparing a book on the recombinant DNA debate, has contributed occasional articles to the American monthly periodical *Environment.* These are careful analytical summaries of current policy decisions.

page 31 "He and another . . ." See J. D. Watson, J. Tooze, *The DNA Story: A Documentary History of Gene Cloning* (Freeman, Reading, 1981).

page 33 "An honourable . . ." See J. Beckwith, "Gene expression in bacteria and some concerns about the misuse of science," *Bacteriological Reviews,* 34 (1970), 222–7. The editor of *Nature,* John Maddox, later to become a public interest representative on the regulatory body concerned with genetic manipulation (GMAG; see text below), denounced these activities from the pages of *Nature,* to which Beckwith and colleagues replied: see "On which side are the angels?" *Nature,* 224 (27 December 1969), 1241–2.

page 33 "This view . . ." The proceedings of this conference were published as W. Fuller (ed.), *The Social Impact of Modern Biology* (Routledge and Kegan Paul, London, 1971).

page 36 "This call..." See P. Berg, D. Baltimore, S. N. Cohen, R. W. Davis, D. S. Hogness, D. Nathans, R. Roblin, J. D. Watson, S. Weissman, N. D. Zinder, "Potential biohazards of recombinant DNA molecules," *Science*, 185 (26 July 1974), 303.

page 38 *"The Ashby Report..."* See *Report of the Working Party on the Experimental Manipulation of the Genetic Composition of Microorganisms* (HMSO, London, 1975).

page 39 "Michael Rogers' article..." See "Pandora's Box congress," *Rolling Stone* (19 June 1975), 15–19, 37–8.

page 39 "The legacy of Asilomar..." See P. Berg, D. Baltimore, S. Brenner, R. O. Roblin, M. F. Singer (Organising Committee of the Asilomar Conference), "Asilomar conference on DNA recombinant molecules," *Nature*, 255 (5 June 1975), 442–4.

page 41 "By the autumn..." The British guidelines for laboratory practice were laid down in the so-called Williams Report, published in August 1976; see *Report of the Working Party on the Practice of Genetic Manipulation* (HMSO, London, Cmnd 6600, 1976).

page 41 "The British committee . . ." The origins and form of the two committees, in the US and the UK, are carefully discussed in S. Wright, "Molecular politics in Britain and the United States: the development of policy for recombinant DNA research," *Southern California Law Review*, 51 (1978), 1383–484.

page 41 "In the mid 1970s . . ." See "Smallpox: ignorance is never bliss," *Nature*, 277 (11 January 1979) 75–6: and also further information on the incident at Birmingham, ibid., 77–81.

page 42 "His report..." See "It takes a death...," *Medical World* (January 1979), 3–13.

page 43 "This has turned out . . ." See the articles in *Medical World* (November–December 1978), 9–15, by Donna Haber, Jerry Ravetz and Bob Williamson. At a Biochemical Society meeting at University College, London, social and political questions were discussed; see Robert Freedman, "Gene manipulation: a new climate," *New Scientist*, 79 (27 July 1978), 268–9.

page 43 "When this issue . . ." See *Neukombinationen von Genen* (Battelle Institut, Frankfurt, 1980).

page 44 "In 1979, four . . ." The proceedings of this meeting were published as J. Morgan, W. J. Whelan (eds), *Recombinant DNA and Genetic Experimentation* (Pergamon Press, Oxford, 1979).

page 46 "As Bob Pritchard . . ." This remark is reported in R. Lewin, "Science and politics in genetic engineering," *New Scientist*, 82 (12 April 1979), 114–15. "The other faction . . ." See E. Lawrence, "Guidelines should go, DNA meeting concludes," *Nature*, 278 (12 April 1979), 590–1.

page 46 "On GMAG's sixth birthday . . ." See Janis Denselow, "GMAG and the teenage jackass," *New Scientist,* 95 (26 August 1982), 558–61.

page 47 "Just what . . ." See John Maddox, "A way to safety without brakes," *Financial Times* (23 May 1975), 8; also Ravetz, op. cit. (see note to page 57 para 4 above).

page 47 "At Wye College . . ." For further thoughts on this meeting see E. J. Yoxen, "Playing God," *Radical Science Journal,* No. 10 (1980), 75–84.

page 48 "These concerns . . ." For a description of the San Francisco company Genentech, see Roger Lewin, "Profile of a genetic engineer," *New Scientist,* 79 (28 September 1978), 924–5.

page 49 "This form of research activity . . ." See *Time* magazine for 9 March 1981, which features Genentech's Herbert Boyer on the cover under the headline, "Shaping life in the lab." The article, "Tampering with beans and genes," *Time* (19 October 1981), discusses the plant genetic engineering companies. That by Glyn Jones, Catherine Bennett, "New genie of genetics," *Sunday Telegraph* magazine (3 October 1982), discusses some of the UK companies, and carries on its cover a picture of Dr. Gerard Fairtlough of Celltech, holding two pieces of an enormous jigsaw. On Transgène see P. Newmark, "France: biotechnology company planned," *Nature,* 285 (15 May 1980), 124.

page 49 "One scientist . . ." See Colin Norman, "First casualty in the biotechnology derby," *Science,* 213 (4 September 1981), 1087–9. DNA Science, the creation of Wall Street stockbrokers E. F. Hutton, was to be a holding company for three ventures, an enterprise called Taglit in Israel, a joint venture with the Battelle Institute in Ohio and Baxter Laboratories in San Francisco. With a target of $40 million, E. F. Hutton raised $8 million themselves, presumably from their clients, two million dollars from nonindustrial sources. The remainder was to come from Allied Chemical and the drug firm Johnson and Johnson, but their terms were too demanding. DNA Science is now being recast as a way of creating tax shelters, see "Bio-taxology," *The Economist* (19 September 1981), 95–6. The new company will be called California Biotechnology Inc., to be based in the San Francisco area: see Deborah Shapley, "Re-entry plans," *Nature,* 298 (19 August 1982), 700.

page 50 "There is real anxiety . . ." See Stephanie Yanchinski, "Universities take to the marketplace," *New Scientist,* 92 (3 December 1981), 675–7.

page 51 ". . . Ray Valentine" See David Dickson, "Conflict of interest on California campus," *Nature,* 293 (8 October 1981), 417.

page 51 "A recent editorial . . ." See "Should academics make money outside?," *Nature,* 286 (24 July 1980), 319–20.

page 52 "An example . . ." see Nicholas Wade, "University and drug firm battle over billion-dollar gene," *Science,* 209 (26 September 1980), 1492–4.

page 53 "As Mrs. Thatcher put it . . ." See David Fishlock, "The biotechnologists Mrs. Thatcher visited," *Financial Times* (12 April 1981), 13. This article describes the growth of Genex, which has now been floated as a public company: see D. Shapley, "Cool reception for Genex shares," *Nature,* 299 (14 October 1982), 573.

page 53 "The Genetics Institute . . ." See Terry Davidson, "DNA firm ready to move into Beacon Street site," *Somerville Journal* (1 January 1981), 1. This article describes some of the controversies in the Greater Boston area over where Ptashne's company would be based; see also Ellen Cantorow, "Gene splicing—not in *our* town," *The Nation* (2–9 January 1982), 14–15.

page 53 "Lord Rothschild . . ." See David Fishlock, "How to pick the genetic company winner," *Financial Times* (13 November 1981), 8. After the first year, $16 million invested in biotechnology had earned only $38,000. The companies that have passed the stringent Rothschild criteria include Agrigenetics of Colorado, which received $1.17 million; Repligen, founded by Drs. Alex Rich and Paul Schimmel of MIT, which got $2.01 million; Genetics Systems of Seattle, which works on monoclonal antibodies; Applied Biosystems, run by professors from Caltech and Colorado, who plan to make gene synthesisers; and Applied Molecular Genetics of Los Angeles: see also "Banker's hope," *Nature,* 299 (7 October 1982), 480.

page 53 "Stockbrokers McNally Montgomery . . ." See Peter Sherlock, "At last—the test tube tax shelter," *Sunday Times Business News* (4 April 1982), 46.

page 54 (quotation) "It begins . . ." From John E. Donalds, "Biotechnology: an American View," *Chemistry and Industry* (7 August 1982), 529.

page 54 "Southern Biotech . . ." See Colin Norman, "Southern Biotech goes bankrupt," *Science,* 216 (18 June 1982), 1297; David Dickson, "Clouds on biotechnology horizon," *Nature,* 296 (4 March 1982), 3.

page 54 ". . . Biogen and Genentech . . ." See Spyros Andreopoulos, "Gene cloning by press conference," *New England Journal of Medicine,* 302 (27 March 1980), 743–6.

page 54 "The price of stock . . ." See David Dugan, "The new biology–a licence to breed money," *The Listener,* 105 (14 May 1981), 633–4.

page 55 para 4. See Mark Cantley, Ken Sargeant, *Strategic Issues for Europe, in the Long-Term Development and Potential Application of Biotechnology* (Commission of the European Communities, Brussels, 1982: draft document No. X11–00482). This document is a part of a study of "The Bio-Society" by the FAST team of Division

X11 of the European Commission, which is largely an exercise in technological forecasting, economic assessment and political analysis from a European perspective.

page 55 "In a sense . . ." See Nicholas Wade, "Court says lab-made life can be patented," *Science,* 208 (27 June 1980), 1445.

page 56 ". . . The actual tasks . . ." see "Gene blues," *Time* (15 June 1981); and David Dickson, "Mixed welcome for genes on Wall Street," *Nature,* 290 (12 March 1981), 77.

page 56 "In the summer of 1982 . . ." A very shrewd appraisal of the possible reasons behind the shake out; see Jeffrey L. Fox, "Biotechnology: a high-stakes industry in flux," *Chemistry and Industry* (29 March 1982), 10–15.

CHAPTER 3

page 61 "In the chromosomes . . ." W. R. Bauer, F. H. C. Crick, J. H. White, "Supercoiled DNA," *Scientific American,* 243 (July 1980), 100–13; A. MacDermott, "Beyond the double helix," *New Scientist,* 95 (22 July 1982), 228–37.

page 65 "In 1977 . . ." See R. Lewin, "Biggest challenge since the double helix," *Science,* 212 (1981), 28–32.

page 69 ". . . a provocative semiobituary . . ." See G. S. Stent, 'That was the molecular biology that was," *Science,* 160 (1968), 390–5.

page 69 "The end of . . ." See F. H. C. Crick, "The genetic code," *Proceedings of the Royal Society of London,* Series B, 167 (1967), 331–47.

page 70 "By the mid 1970s . . ." The technical details of this are well covered in Jeremy Cherfas' book (see the note to chapter 1, page 9, para 4 above) and in chapter 16 of Watson and Tooze, op. cit. (see the note to chapter 2, page 31, para 3).

page 73 "Here then . . ." See S. N. Cohen, "The manipulation of genes," *Scientific American,* 233 (July 1975), 25–33.

page 76 "The plan . . ." See Colin Norman, "Genetic manipulation to be patented?" *Nature,* 261 (24 June 1976), 624.

page 76 "Helling was not pleased . . ." See David Dickson, "Inventorship dispute stalls DNA patent application," *Nature,* 284 (3 April 1980), 388.

page 77 "They decided . . ." See David Dickson, "Stanford ready to fight for patent," *Nature,* 292 (13 August 1981), 573.

page 77 "At the beginning . . ." See Stephen Budiansky, "More trouble for Stanford's prize," *Nature,* 298 (12 August 1982), 595.

page 77 ". . . prior publication . . ." The article referred to here is E. Ziff, "Benefits and hazards of manipulating DNA," *New Scientist,* 60 (25

October 1973), 274–5. This was an account of the Gordon conference at which Berg's proposed experiments produced concern.

page 77 "The earlier patent . . ." See Stephen Budiansky, "Key biotechnology patent delayed," *Nature*, 298 (29 July 1982), 409–10; and Paula Dwyer, "Genetic engineers stitched up over patent rights," *New Scientist*, 95 (8 July 1982), 75; also S. Budiansky, "Further snag with Stanford patent," *Nature*, 300 (11 November 1982), 95–6; S. Budiansky, "Biotechnology patent challenged: ex-colleague seeks share of the credit," *Nature*, 300 (25 November 1982), 303.

page 78 "Mere secretive tinkering . . ." For an argument against the patenting of modified life forms as such, and patenting in this area of research in general, see Jonathan King, "The case against . . . ," *Environment*, 24 (July/August 1982), 38–9. The opposite position is put there by Charles E. Lipsey and Carol P. Einaudi, ibid., 39–41.

page 79 Quotation from Chief Justice Burger, "From majority opinion," published in the *New York Times* (17 June 1980), 17.

page 79 "When some years ago . . ." See Stephen Crespi, "Patenting nature's secrets and protecting microbiologists' interests," *Nature*, 284 (17 April 1980), 590. The entire issue of the *University of Toledo Law Review*, 12, No. 4 (1981), is given over to a comparative discussion of patenting in various countries. The publishers, Macmillan, have just produced a reference book for the biotechnology industry, *Genetic Engineering Patents*, 1980–81. See also D. W. Plant, N. J. Reimers, N. D. Zinder (eds), *Patenting of Life Forms* (Cold Spring Harbor Laboratory, New York, 1982).

page 79 "Companies may rely . . ." See Barry Fox, "Is the day of the patent over?" *New Scientist*, 91 (10 September 1981), 633–5.

page 81 ". . . gene machines . . ." See "The seamier side of gene machines," *New Scientist*, 89 (29 January 1981), 261.

page 81 "The most striking . . ." See "The longest synthetic gene . . . so far," *Nature*, 292 (20 August 1981), 667; also M. D. Edge et. al., "Total synthesis of human leukocyte interferon gene," ibid., 756–62.

page 82 "This confident approach . . ." See "With computers," *The Economist*, 284 (11 September 1982), 83–4; "Venture weds computer and biotechnologies," *Chemical Engineering News* (31 May 1982).

page 82 "In Japan . . ." See M. D. Rogers, "The Japanese government's role in biotechnology R and D," *Chemistry and Industry* (7 August 1982), 533–7.

page 83 "The yield . . ." See Y. Aharonowitz, G. Cohen, "The microbiological production of pharmaceuticals," *Scientific American*, 245 (September 1981), 106–19 at page 117.

page 83 "With that change . . ." See U. Faust, "Engineering aspects in biotechnology," *Chemistry and Industry* (7 August 1982), 527–8; see also J. E. Smith, *Biotechnology* (Edward Arnold, London, 1981).

page 85 "These are called . . ." See D. Thomas, T. Gellf, "Enzyme engineering: accomplishments and prospects," *Endeavour*, 5, No. 3 (1981), 96–8.

page 85 "The business . . ." See R. Sherwood, T. Atkinson, "New microbes for genetic engineers," *New Scientist*, 91 (10 September 1981), 665–7.

CHAPTER 4

page 88 "Corporations which . . ." For a discussion of how and why this threefold relationship of power has come about see Lesley Doyal, op. cit. (see the note to page 3 para 2).

page 89 "However the honour . . ." See Schuuring, op. cit. The marketing of insulin is described in P. Newmark, "Insulin on tap," *Nature*, 299 (23 September 1982), 293.

page 91 "The rate of increase . . ." See *A Study of Insulin Supply and Demand: A Report of the National Diabetes Advisory Board* (US Government Printing Office: Department of Health, Education and Welfare, Publication No. (NIH) 78–1588).

page 91 "Since the 1950s . . ." See M. Silverman, P. R. Lee, *Pills, Profits and Politics* (University of California Press, London, 1974).

page 92 "In September 1982 . . ." See J. Ehrlichman, "Biochemists go to war —and UK is the battleground," *The Guardian* (15 September 1982).

page 92 "The price . . ." See Ehrlichman, op.cit., and Newmark, op.cit.

page 93 "This model . . ." See H. C. Trowell, "Hypertension, obesity, diabetes mellitus and coronary heart disease," in H. C. Trowell, D. P. Burkitt (eds), *Western Diseases: Their Emergence and Prevention* (Edward Arnold, London, 1981), 3–32.

page 95 "A few investigators . . . " See K. Cantell, "Why is interferon not in clinical use today?" in I. Gresser (ed.), *Interferon* (Academic Press, London, 1979), 2–27.

page 96 "In January 1980 . . ." See Andreopoulos, op. cit. (see the note to page 70 para 5 above).

page 96 "What are we to make . . ." See G. M. Scott, D. A. J. Tyrell, "Interferon: therapeutic fact or fiction for the 80s," *British Medical Journal* (28 June 1970), 1558–62; F. Balkwill, "What future for the interferons?" *New Scientist*, 84 (24 January 1980), 230–2; "What not to say about interferon" (editorial), *Nature*, 285 (26 June 1980), 603–4; P. Newmark, "Interferon: decline and stall," *Nature*, 291 (14

May 1981), 105–6; R. M. Friedman et al., "Interferon redux," *Nature*, 296 (22 April 1982), 704–5; Robert Walgate, "Interferon therapy side effect scare hits French trials," *Nature*, 300 (11 November 1982), 97–8.

page 97 "There are reports . . ." See T. Boddé, "Interferon: will it live up to its promise?" *BioScience*, 32, No. 1 (1982), 13–15; R. S. Johnson, "Interferon: cloudy but intriguing future," *Journal of the American Medical Association*, 245 (9 January 1981), 109–10, 115–16.

page 98 "In America . . ." See E. R. Gonzalez, "Teams vie in synthetic production of human growth hormone," *Journal of the American Medical Association*, 242 (24–31 August 1979), 701–2.

page 98 "At present, the hormone . . ." See R. Walgate, "Pituitary slump," *Nature*, 290 (5 March 1981), 6–7.

page 99 "Certainly, one regional . . . " See "Glands taken illegally," *The Guardian* (31 October 1981).

page 99 "In 1982 . . ." See B. Aberg, "L'hormone de croissance humaine par génie génétique," *Biofutur*, No. 4 (juin 1982), 29–33.

page 100 "In 1982, the business . . ." See T. Leeney, "Health care applications," *Chemistry and Industry* (7 August 1982), 518–21.

page 100 "One estimate . . . " ibid.

page 101 "In Japan . . ." See R. Whymant, "The brutal truth about Japan," *The Guardian* (14 August 1982), 15.

page 101 "The technology . . ." See D. Bevan, M. Rose, "Perfidious albumin: the politics of plasmaphoresis," *World Medicine* (13 June 1981), 73–6.

page 103 "The immune response . . ." See chapter 19, "The problem of antibody synthesis," in J. D. Watson, op. cit. (see the note to page 19 para 2).

page 104 "This was first done . . ." See C. Milstein, "Monoclonal antibodies," *Scientific American*, 243 (October 1980), 56–64; and G. Kohler, C. Milstein, "Continuous cultures of fused cells secreting antibody of predefined specificity," *Nature*, 256 (1975), 495–7.

page 104 "This work . . ." See M. D. Scharff, S. Roberts, "Present status and future prospects of hybridoma technology," *In vitro*, 17 (December 1980), 1072–7; O. Gillie, "Medical advance offers hope of 'magic bullet,' " *Sunday Times*, (24 May 1981), 3.

page 105 "This would allow . . ." See J. Hanmer," Sex predetermination, artificial insemination and the maintenance of a male-dominated culture," in H. Roberts (ed.), *Women, Health and Reproduction* (Routledge and Kegan Paul, London, 1981), 163–90.

page 105 "One estimate . . ." See "Biotechnology's new thrust into antibodies," *Business Week* (18 May 1981), 147.

page 106 "One attempt . . ." See R. Walgate, "British gene company opens books," *Nature*, 289 (26 February 1981); D. Fishlock, "Celltech applies British genius to British genes," *Financial Times* (9 October 1981), 15; R. McKie, "Dispensers of knowledge at £400 a millilitre," *The Times Higher Education Supplement* (18 December 1981), 9.

page 107 "In the spring . . ." See *Biotechnology: Interim Report on the Protection of the Research Base in Biotechnology* (Sixth Report from the Education, Science and Arts Committee, Session 1981-2), (HMSO, London, 1982), pp. xxx–xxxi.

page 108 "By 1980 . . ." See N. Wade, "La Jolla biologists troubled by the Midas factor," *Science*, 213 (7 August 1981), 623–8; see also S. Budiansky, "Mass. General placates Hoechst," *Nature*, 300 (25 November 1982), 305. This describes a case where Massachusetts General Hospital in Boston, having secured $50 million from the Hoechst Chemical Company, had to take legal steps to ensure that the existing consultancy arrangements of one of the MGH molecular biologists, Brian Seed, with Genetics Institute—another biotechnology company—would be acceptable to Hoechst. The different interests of Seed, MGH, Hoechst and Genetics Institute had to be reconciled by negotiations between several firms of lawyers, over what Seed should be allowed to say, to whom, and when.

page 110 "The technique of . . ." See T. M. Maugh, "FDA approves hepatitis vaccine," *Science*, 214 (4 December 1981), 1113; J. Redfearn, "Race to vaccine," *Nature*, 298 (15 July 1982), 218; "Vaccines: Japanese development," *Financial Times* (24 August 1982), 7; P. D. Minor et. al., "Poliomyelitis-epidemiology, molecular biology and immunology," *Nature*, 299 (9 September 1982), 109–10.

page 110 "One of the most . . ." See H. Vrbova, P. Heywood, "Malaria: prevention or cure?" *Medicine in Society*, 8, No. 1 (1982), 40–2; "Malaria vaccine?," *Africa Health* (December 1981–January 1982), 37; "Malaria vaccine engineered genetically," *New Scientist*, 88 (15 January 1981), 131.

page 111 "What is striking . . ." See W. F. Anderson, J. C. Fletcher, "Gene therapy in human beings: when is it ethical to begin?" *New England Journal of Medicine*, 303 (27 November 1980), 1293–6; "The promise of gene therapy," *New Scientist*, 93 (21 January 1982), 164; R. Williamson, "Gene therapy," *Nature*, 298 (29 July 1982), 416–18.

page 113 "Some American states . . ." See P. Reilly, *Genetics, the Law and Social Policy* (Harvard University Press, Cambridge, Mass., 1978).

page 114 "Those setting up . . ." See E. J. Yoxen, "Constructing genetic diseases" in P. Wright, A. Treacher (eds), *The Problem of Medical Knowledge: Towards a Social Constructivist View* (Edinburgh University Press, Edinburgh, 1982), 144–61.

page 114 "The successful attempts . . ." See the note to page 8 para 4.

page 115 "By 1980 . . ." M. J. Cline et al., "Gene transfer in intact animals," *Nature,* 284 (3 April 1980), 422–5; K. E. Mercola, M. J. Cline, "The potentials of inserting new genetic information," *New England Journal of Medicine,* 303 (27 November 1980), 1297–300.

page 115 "When the details . . ." See "Genetic engineer who broke the rules is punished—a little," *New Scientist,* 90 (4 June 1981), 605.

CHAPTER 5

page 119 "The yields of . . ." See "The second green revolution," *Business Week* (25 August 1980), 2–8. Another striking statistic is that the average US yield of corn by the mid 1970s was 5.8 tons per hectare. Of the 135 nations where corn is grown, 112 had average yields of under 3 tons per hectare and 81 of them less than 1.5 tons per hectare. Fertiliser is by no means the only essential factor. Water is also vital. It takes 500,000 gallons of water to grow an acre of rice.

page 119 "It has been calculated . . ." See C. C. Delwiche, "Legumes—past, present and future," *BioScience,* 28, No. 9 (September 1978), 565–70; V. P. Gutschick, "Energy and nitrogen fixation," ibid., 571–5.

page 120 "One estimate . . ." See N. Myers, "Plants that are worth their salt," *The Guardian* (4 March 1982), 22.

page 121 "The pattern . . ." See J. Walsh, "Biotechnology boom reaches agriculture," *Science,* 213 (18 September 1981), 1339–41.

page 122 "Some of the new plants . . ." See "Tampering with beans and genes," *Time* (19 October 1981); A. Veitch, "Biting back at the bug," *The Guardian* (17 May 1982).

page 122 "An important precondition . . ." See A. D. Krikorian, "Cloning higher plants from aseptically cultured tissues and cells," *Biological Reviews of the Cambridge Philosophical Society,* 57 (May 1982), 151–218; T. A. Thorpe, *Plant Tissue Culture: Methods and Applications in Agriculture* (Academic Press, New York, 1981); I. K. Vasil et al., "Plant tissue culture in genetics and breeding," *Advances in Genetics,* 20 (1974), 127–215.

page 123 "One example . . ." This is discussed in the *Business Week* article mentioned above; see also J. F. Shepard, "The regeneration of potato plants from leaf cell protoplasts," *Scientific American,* 246 (May 1982), 154–66.

page 124 "Nitrogen is the key . . ." See W. J. Brill, "Biological nitrogen fixation," *Scientific American,* 236 (March 1977), 68–81.

page 125 "Between 1950 . . ." See Delwiche, op. cit. (see the note to page 145 para 1 above).

page 126 "One option . . ." See W. J. Brill, "Agricultural microbiology," *Scientific American,* 245 (September 1981), 146–56.

page 127 "UNESCO is . . ." See E. J. DaSilva, "Les banques de souches," *Biofutur*, No. 6 (octobre 1982), 35–40. This article describes the establishment of Microbiological Resources Centres (MIRCENS) in Brisbane, Bangkok, Guatemala, Stockholm, Cairo, Porto Alegre (Brazil), Nairobi and Pahaia (Hawaii).

page 127 "Something like this . . ." See G. A. Peters, "Blue-green algae and algal associations," *BioScience*, 28, No. 9 (September 1978), 580–5.

page 128 "The obvious implication . . ." This timescale is supported by questionnaire data obtained from plant breeders relating specifically to maize; see K. M. Menz, C. F. Neumeyer, "Evaluation of five emerging biotechnologies for maize," *Bio-Science*, 32, No. 8 (September 1982), 675–6.

page 129 "From the point of view . . ." This tightening correspondence between chemical packages or 'total plant systems' and specific plant varieties is discussed in detail in M. Kenney et al., *Genetic Engineering and Agriculture: Exploring the Impacts of Biotechnology on Industrial Structure, Industry-University Relationships and the Social Organisation of US Agriculture* (Bulletin No. 125, Department of Rural Sociology, Cornell University Agriculture Experiment Station, New York 14853, July 1982).

page 130 "One report . . ." See "Unilever's World," *CIS Anti-Report*, No. 11 (Counter Information Services, London, no date).

page 130 "As a former chairman . . ." See the remarks of Lord Cole, reported in the *Financial Times*, (28 May 1966) and cited in "Unilever's World," page 93. The continuing importance of R&D to the company is described in Sir David Orr, *Research and Development in Unilever* (Unilever Public Relations Department, London, 1981).

page 130 "In 1968 . . ." See L. Jones, *Clonal Oil Palm Propagation by Tissue Culture* (Unilever Research Laboratory, Sharnbrook, Bedfordshire, no date).

page 131 "In 1978 a rough . . ." See *Application of Biotechnology to the Food Industry: An Appraisal* (Food Research Association, Leatherhead, Surrey, 1981), 7.

page 131 "The American timber . . ." See J. Elkington, "Test tube trees," *The Guardian* (28 January 1982), 14.

page 132 "The most serious consequence . . ." See "The vanishing jungle," *The Economist* (4 September 1982), 89–90.

page 132 "They are vital . . ." See N. Myers, "What gets the chop when the axe falls," *The Guardian* (10 December 1981).

page 134 "Less well known . . ." See G. E. Seidel, "Superovulation and embryo transfer in cattle," *Science*, 211 (23 January 1981), 351–8; A.

Crittenden, "A breed apart; building a better cow," *New York Times* (22 March 1981), F7.

page 136 "For the moment . . ." For developments in the culture of other animals see, for example, "How about a battery lobster?" *The Economist* (4 September 1982), 94.

page 136 "This question . . ." See J. Vandermeer, "Tomatoes in the Midwest: Agricultural research and social conflict," *Science for the People* (January–February 1981), 5–8, 25–30.

page 139 "For example . . ." See D. Dickson, "University challenged over 'agribusiness' connections," *Nature*, 278 (26 April 1979), 768–9.

page 141 "Trofim Denisovitch Lysenko . . ." See R. C. Lewontin, R. Levins, "The Problem of Lysenkoism," in H. Rose, S. Rose (eds), *The Radicalisation of Science* (Macmillan, London, 1976), 32–64; R. M. Young, "Getting started on Lysenkoism," *Radical Science Journal*, No. 6/7 (1978), 81–106.

page 142 "By the mid 1960s . . ." See N. Wade, "Green revolution (I): a just technology, often unjust in use," *Science*, 186 (20 December 1974), 1093–6.

page 143 "Supporters of the . . ." See, for example, D. L. Plucknett, N. J. H. Smith, "Agricultural research and Third World food production," *Science*, 217 (16 July 1982), 215–20.

page 144 "The control of cash . . ." See B. Roy, "Poverty keeps the rich in power," *The Guardian* (8 October 1982), 15.

page 144 "In agricultural terms . . ." See N. Wade, "Green revolution (II): problems of adapting a Western technology," *Science*, 186 (27 December 1974), 1186–92.

page 145 "To sustain . . ." For a comprehensive discussion of all these issues see A. Pearse, *Seeds of Plenty, Seeds of Want: Social and Economic Implications of the Green Revolution* (Oxford University Press, London, 1980).

page 145 "In a letter . . ." See P. R. Mooney, *Seeds of the Earth: A Private or Public Resource* (International Coalition for Development Action, London, 1979). This is available from ICDA, Bedford Chambers, Covent Garden, London WC2.

page 145 "Mooney and others . . ." See also C. Fowler, "Plant patenting: sowing the seeds of destruction," *Science for the People* (September–October 1980), 8–10; C. Fowler, "Testimony on behalf of the National Sharecroppers' Fund before the Senate Agriculture Subcommittee on Agricultural Research and General Legislation on S. 23, an amendment to the Plant Variety Protection Act, June 17, 1980." This is available from the National Sharecroppers' Fund, Frank Parker Graham Center, Rt. 3, Box 95, Wadesboro, North

Carolina 28170 USA. See also *Plant Variety Protection Act Amendments: Hearings before the Subcommittee on Departmental Investigations, Oversight and Research of the Committee on Agriculture, House of Representatives, 96th Congress, 1st and 2nd Sessions on H.R. 999, Senate No. 96–CCC, April 22 and July 19 1980* (US Government Printing Office, Washington, DC, 1980).

page 146 "Thus in 1970 . . ." See D. H. Smith, J. King, "The legislative and legal background," *Environment*, 24, No. 6 (July–August 1982), 24–26; see also V. Sarma, "Australian patents bill: seeds of doubt," *Nature*, 298 (26 August 1982), on the dispute in Australia on the Plant Variety Rights Bill which would allow the kind of proprietary restriction already offered to plant breeders in Europe and America.

page 148 "Without a wide range . . ." See R. and C. Prescott-Allen, *Wild Plants and Crop Improvements* (World Wildlife Fund, Godalming, 1981); see also, by the same authors, "Reaping the wild oats," *The Guardian* (10 September 1981), 15.

page 148 "Pioneer Hi-Brid . . ." See J. Walsh, "Germ plasm resources are losing ground," *Science*, 214 (23 October 1981), 421–3.

page 149 "When I wrote . . ." The brochure that I received is *Feeding the 5000 million* (ASSINSEL, Amsterdam, no date). This is available from ASSINSEL, Rokin 50, 1012 KV, Amsterdam, Holland.

CHAPTER 6

page 151 "In the 1950s . . ." See D. Beynon, "Rationalisation: an ICI view," *Chemistry and Industry* (16 January 1982), 46–9.

page 151 "Four factors . . ." S. Ahearne, "Petrochemicals: is there a future?" ibid., 42–6.

page 151 "With current technology . . ." The unit cost of ethylene falls from £220 per ton to £170 per ton as the size of the ethylene cracker is increased from 200,000 tons per year to 500,000 per year; see Ahearne, op. cit., figure 2, page 43.

page 152 "All this can . . ." See T. A. Sparta, "Europe after change," *Chemistry and Industry* (16 January 1982), 52–5.

page 153 ". . . the role of R&D . . ." See W. F. Fallwell, "Company R and D budgets set for robust 1982," *Chemical and Engineering News* (18 January 1982), 52–5. This article, based on a survey of the US basic chemical industry by the Battelle Laboratories in Columbus, Ohio, shows that the R&D spending by the US basic chemical industry now stands at $2600 million per year, which is three per cent of sales, and that 24,000 full-time equivalent R&D scientists and engineers are employed by this industry.

page 153 "That, as it happens . . ." See Elkington, op. cit. (see the note on page 2 para 4), The Fox, op. cit. (see the note to page 56 para 2), on the questionable value of some of the surveys. The newsletters are reviewed in J. Davies, "What's news in biotechnology?" *Nature,* 299 (7 October 1982), 493–6. The annual subscriptions range from between £ 50 and £ 500. The going rate for business briefing conferences is remarkably high, of the order of £100 to £500 per day for each participant. For that you can hear a range of talks on biotechnology and its business aspects, varying in quality from the excellent to the laughable, and a chance to meet your present or future competitors. Technology forecasting is at best an uncertain game. Two different groups, one at Genex, the other at MIT, produced remarkably different scenarios for the chemical industry in an exercise commissioned by the Office of Technology Assessment in the United States. Out of a list of fifty-seven chemicals they agreed only in nine cases as to whether the substance concerned would be made biologically within twenty years. In fairness it should be said that these groups used different assumptions and asked slightly different questions: see "What applied genetics might do in chemicals," *Chemical Week* (4 March 1981), 41–2.

page 154 "How then does . . ." This is largely drawn from P. G. Caudle, "The chemical industry—future energy and feedstock needs," *Chemistry in Britain* (April 1982), 256–60.

page 154 "In this situation . . ." See "Designing chemical plants to save energy," *The Economist* (6 June 1981), 89.

page 154 "Broadly speaking . . ." See R. M. Ringwald, "Energy and the chemical industry," *Chemistry and Industry* (1 May 1982), 281–6; see especially tables 4, 5 and 6.

page 154 "First, one strategy . . ." See, for example, Sparta, op. cit. (see the note to page 152 para 4).

page 155 "One strategic option . . ." See "Petrochemicals from coal," *The Economist* (1 August 1982), 66–7.

page 155 "Methanol looks like . . ." See M. E. Frank, "Methanol: emerging uses, new syntheses," *Chemtech* (June 1982), 358–62; J. Haggin, "Methanol from biomass draws closer to market," *Chemical and Engineering News* (12 July 1982), 24–5; K. A. Kovaly, "Biomass chemicals," *Chemtech* (August 1982), 486–9.

page 155 "At the moment . . ." See J. H. Krieger, "Achema 82 charts chemical technology trends," *Chemical and Engineering News* (21 June 1982), 26–30; and "Progress slows in chemicals from synthetic feeds," *Chemical and Engineering News* (29 March 82), 28.

page 158 "papain . . ." See D. Eveleigh, "The microbiological production of industrial chemicals," *Scientific American,* 245 (September 1981), 120–30. Worldwide sales of enzymes amounted to $300 million in

1980, of which $66 million went on detergents and $64 million for cheese-making.

page 158 "It has recently . . ." See "Gene cloned for enzyme used to make cheese," *Chemical and Engineering News* (15 February 1982), 34.

page 159 "Enzymes can be bound . . ." See Thomas, Gellf, op. cit. (see the note to page 85 para 2).

page 159 "But a route . . ." See Eveleigh, op. cit. (see the note to page 158 para 2).

page 160 "One can't help . . ." See D. Shapley, "Cetus goes begging," *Nature*, 297 (24 June 1982), 616.

page 160 "Rather it has spurred . . ." See "New genes promise a sweet future for cassava," *New Scientist*, 91 (3 September 1981), 599.

page 161 "Let us assume . . ." See Eveleigh, op. cit., 124 (see the note to page 158 para 2).

page 161 "It is worth . . ." See D. Pimentel et al., "Biomass energy from crop and forest residues," *Science*, 212 (5 June 1981), 1110–15.

page 162 "Lactic acid . . ." See E. S. Lipinsky, "Chemicals from biomass: Petrochemicals substitution options," *Science*, 212 (26 June 1981), 1465–71.

page 163 "There is a nice story . . ." See D. Fishlock, "Exxon looks for enzyme factory," *Financial Times* (27 November 1981), 14.

page 163 "Personal transportation . . ." See A. Spinks, "Alternatives to fossil petrol," *Chemistry in Britain* (February 1982), 99–105.

page 164 "This tree . . ." See M. Calvin, "Petroleum plantations for fuel and materials," *BioScience*, 29, No. 9 (September 1979), 533–8.

page 164 "Ernest Bungay . . ." See H. R. Bungay, *Energy, The Biomass Options* (Wiley, New York, 1981).

page 165 "One estimate . . ." See N. Myers, "The four-star harvest," *The Guardian* (1 July 1982), 17.

page 166 "The latex . . ." See P. James, "Reagan's spare tyre shrub," *The Guardian* (14 January 1982), 13.

page 166 "In seeds . . ." See D. O. Hall, "Put a sunflower in your tank," *New Scientist*, 89 (26 February 1982), 524–6.

page 168 "Brazil's foreign debt . . ." See M. Bazin, "Brazil: running on alcohol," *Nature*, 282 (6 December 1979), 550–2; see also H. Rothman, F. Rosillo-Calle, R. Greenshields, *The Alcohol Economy: Fuel Ethanol and the Brazilian Experiment* (Frances Pinter, London, 1982).

page 169 "One ICI commentator . . ." See Spinks, op. cit. (see the note to page 163 para 3).

page 169 "If you assume . . ." See Spinks, op. cit. Interestingly the figures for the proportion of total land area required vary from 23% (Eire) to 350% (Benelux). These figures assume a production of twelve tons dry weight of biomass per hectare per year.

page 170 "Organic effluents . . ." See A. D. Wheatley, "Effluent treatment," *Chemistry and Industry* (7 August 1982), 512–18.

page 171 "Waste processing . . ." See also Smith, op. cit., chapter 9, "Environmental technologies" (see the note to page 83 para 2); see also J. Haggin, "Bacterium may clean industrial gas streams," *Chemical and Engineering News* (10 May 1982), 48–9.

page 171 "One can confidently . . ." See B. Kovarik, *Fuel alcohol; energy and environment in a hungry world* (Earthscan, London, no date).

CHAPTER 7

page 178 ". . . flotations in Britain . . ." See A. Coghlan, "Biotechnology—the Welsh whey forward," *Chemistry and Industry* (4 September 1982), 615. This small company processes wastes from cheese-making, to produce high-grade protein. Having battled for seven years to get development funding, its share offer was oversubscribed by fifteen times. Since I wrote this the share price of Biolsolates has rocketed, which weakens my case considerably.

page 178 ". . . in the case of DNA Science . . ." See the note to page 49 para 5.

page 178 "The BTG . . ." See "Taxpayer's British Technology," *The Economist* (25 July 1981), 18.

page 179 ". . . at this stage . . ." This argument is put persuasively in R. M. Young, op. cit.

page 180 "There is no doubt . . ." This message is clearly spelt out in the Spinks Report (see the note to page 29 para 6) and in the evidence submitted to the House of Commons Select Committee hearings on Biotechnology (see the note to page 107 para 1), in the aftermath of the government White Paper, which was a response to the Spinks Report: see *Biotechnology* (HMSO, London, 1981). This report was the subject of much bitter discussion at the Second International Congress of Biotechnology, at Eastbourne in the spring of 1981.

page 180 "One of them . . ." See the article by Jones and Bennett (see the note to page 49 para 2); see also "Who's behind, who's ahead?" *Nature*, 283 (10 January 1980), 123–4.

page 181 "All these things . . ." Just how and why the debate differed in the UK from that in the US is discussed in J. W. Turney, *Recombinant DNA: Influences on Public Attitudes to a Scientific Innovation* (Unpublished MSc thesis, Manchester University).

page 181 "... Charles Weiner ..." See C. Weiner, "Science in the market-place: historical precedents and problems," to appear in *From Genetic Experimentation to Biotechnology: The Critical Transition* (Wiley, New York, 1982).

page 182 "An example..." This is based on W. Lepkowski, "Academic values tested by MIT's new center," *Chemical and Engineering News* (15 March 1982), 7–12; see also D. F. Noble, "The selling of the university," *The Nation* (6 February 1982), 143–8; also "MIT on Wall Street," *New Scientist,* 90 (28 May 1981), 542. This describes an agreement between MIT and Flow General Inc. to develop a process originating from an MIT research group that allows human and mammalian cells to be grown on microcarrier beads. MIT stands to gain $400,000 a year from this. Finally, see also "Grace, MIT sign pact on biotechnology research," *Chemical and Engineering News* (9 August 1982), 5. This agreement involves $6–8.5 million over five years. Federally funded research at MIT has grown by ten per cent from 1977 to 1982, from $102 million to $157 million. In this period industrially funded research has expanded from $6.7 million to $20.3 million, that is from six per cent of the total to twelve per cent.

page 185 "For example in the UK ..." See "Now Monsanto," *Nature,* 298 (12 August 1982), 598–9. This fund, loosely connected with Monsanto, will invest £500,000 from Cambridge University, £100,000 from Oxford University and undisclosed sums from Imperial College, London, Boston University (Massachusetts) and the Nuffield Foundation, in a range of new ventures.

page 190 "The arrangement between . . ." See "Chemical giant adds to sponsorship craze," *The Times Higher Education Supplement* (10 July 1981); this refers to two investments, one of $6 million by Du Pont in a new genetics department at Harvard Medical School, and another of $50 million to the Massachusetts General Hospital by Hoechst; see also B. J. Culliton, "Monsanto gives Washington University $23.5 million," *Science,* 216 (18 June 1982), 1295–6.

page 194 "The National Research and . . ." See J. Stansell, "NRDC's first priority—selling itself to Britain," *New Scientist,* 79 (14 September 1979), 760–1; and R. McKie, "Commercial breakdown," *The Times Higher Education Supplement* (22 January 1982), 9. The latter article reports on an academic study of the NRDC, which identifies its relative failure to commercialise the inventions offered to it. An alternative to reorganization of the NRDC would be to privatise it. This kind of company, specialising in technological brokerage, already exists in the UK and elsewhere. An American example is Ugen—or University Genetics of Connecticut—see C. Joyce, "New company could turn academics into tycoons," *New Scientist,* 90 (28 May 1981), 542.

page 194 "one example is WARF . . ." See S. Yanchinski, op. cit. (see the note to page 50 para 4).

page 195 "Mrs Thatcher's visit . . ." See "Thatcher applauds genes," *Nature*, 298 (12 March 1981), 78.

page 195 "Finally we need . . ." See S. Wright, "The status of hazards and controls," *Environment*, 24, No. 6 (July–August 1982), 12–17, 20, 51–3.

Further Reading

Jeremy Cherfas, *Man-Made Life: A Genetic Engineering Primer* (Blackwell, 1982). This is an introduction to the technicalities of gene-splicing and to some of the possible applications.

Industrial Microbiology: the issue of *Scientific American* for September 1981. This is a rather more detailed and technically advanced survey of the field. It covers many of the areas that have been discussed here, though it has nothing to say about the social implications.

James D. Watson, John Tooze (eds), *The DNA Story: A Documentary History of Gene Cloning* (Freeman, 1981). A valuable collection of documents that conveys something of the feel of the recombinant DNA debate, from two very active participants in that debate.

Sheldon Krimsky, *Genetic Alchemy: The Social History of the Recombinant DNA Controversy* (MIT Press, 1982). A more academic treatment of the same material.

Biotechnology: Report of a Joint Working Party (HMSO, 1980). The Spinks report: this is not a document of electrifying interest, but it is a survey of the problems facing biotechnological innovation in the UK as biotechnologists see them.

Index

Acetic acid, 162
Adenine, 63
Advanced Genetics Science Ltd., 123
Advent Management Ltd., 185
African Explosives and Chemical Industry, 155
Agriculture, 12, 118–29, 140–49, 188, 198
  fertilisers in, 119, 124, 125
  See also specific plants and crops
Agrigenetics, 147–48
Agrobacterium tumefaciens, 128
Alcohol, 167
  ethyl (ethanol), 160–62
Aliphatic organic compounds, 160
Allied Chemical Corporation, 51
Alpha-amylase, 158–59
American Cancer Society, 95
American Seed Trade Association, 145
Amino acids, 162–63
  sequence of, 61–66
Anabaena azollae, 127
Andreopoulos, Spyros, 54
Anfinsen, Christian, 49
Animals, 134–36
  feedstuff for, 7
Animal vaccines, 110
Antibiotics, 72
Antibodies
  functions of, 103
  monoclonal, 102, 104–7
  structure of, 103–4
Antigens, 103–4
Apple, Martin, 82
Artificial insemination, 134
Asclepsia speciosa (milkweed), 166
Ashby, Lord, 38

Ashby report (1975), 38
Asilomar conference (1975), 38–39
Aspergillus niger, 162
ASSINSEL, 149
Association of Scientific, Technical and Managerial Staffs (ASTMS), 41–43
Azolla fern, 127

Bacteria, 66–68
  industrial-scale culture of, 83–85
  nitrogen-fixing, 124–29
  thermophilic, 158–59
Baltimore, David, 182–84
Beckwith, Jonathan, 33, 34, 69
Berg, Paul, 35–38, 75
Beta-galactosidase, 80–81
Biogen, 54
Biology, Rockefeller Foundation and, 21–22
Biomass, 163–69
Biotechnology, 1–16
  agenda for, 11–13
  alternative futures of, 13
  constituencies of, 196–98
  examples of achievements and potential of, 3–7
  future of, 198–201
  investment in, 178–81, 186–87, 190, 192–93
  medical, 87–117, 178–88, 197–98
  moving genes around as essence of, 7–9
  as nothing new, 13–14
  as revolution, 9–10, 14, 174–76
  secrecy of research in, 176–77
  See also specific topics
Biotechnology Investments Ltd., 53
Birmingham University, 42
Blood products, 101–2

228 | INDEX